Texts in Computational Science and Engineering

2

Springer
Berlin
Heidelberg
New York
Hong Kong
London
Milan
Paris
Tokyo

Alfio Quarteroni
Fausto Saleri

Scientific Computing
with MATLAB

Springer

Alfio Quarteroni
Ecole Polytechnique Fédérale de Lausanne
1015 Lausanne, Switzerland
and Politecnico di Milano
Via Bonardi 9
20133 Milano, Italy
e-mail: alfio.quarteroni@epfl.ch

Fausto Saleri
Politecnico di Milano
Via Bonardi 9
20133 Milano, Italy
e-mail: fausto.saleri@mate.polimi.it

Cover figure by Marzio Sala

Cataloging-in-Publication Data applied for
Bibliographic information published by Die Deutsche Bibliothek

Die Deutsche Bibliothek lists this publication in the Deutsche Nationalbibliografie;
detailed bibliographic data is available in the Internet at <http://dnb.ddb.de>.

Title of the Italian original edition: *Introduzione al Calcolo Scientifico.*
Springer-Verlag Italia, Milano, 2001, ISBN 88-470-0149-8

Mathematics Subject Classification (2000): 65-01, 68U01, 68N15

ISSN 1611-0994
ISBN 3-540-44363-0 Springer-Verlag Berlin Heidelberg New York

Springer-Verlag Berlin Heidelberg New York
a part of Springer Science+Business Media

http://www.springer.de

© Springer-Verlag Berlin Heidelberg 2003
Printed in Germany

Cover Design: Friedhelm Steinen-Broo, Estudio Calamar, Spain
Cover production: *design & production* GmbH, Heidelberg
Typeset by the authors.

Printed on acid-free paper   46/3111 – 5 4 3 2 1   SPIN 11017271

Introduction

*"It is important to prove,
but it is more
important to improve."*

This textbook is an introduction to Scientific Computing. We will illustrate several numerical methods for the computer solution of certain classes of mathematical problems that cannot be faced by paper and pencil. We will show how to compute the zeros or the integrals of continuous functions, solve linear systems, approximate functions by polynomials and construct accurate approximations for the solution of differential equations.

With this aim, in Chapter 1 we will illustrate the rules of the game that computers adopt when storing and operating with real and complex numbers, vectors and matrices.

In order to make our presentation concrete and appealing we will adopt the programming environment MATLAB $^{\textregistered}$ [1] as a faithful companion. We will gradually discover its principal commands, statements and constructs. We will show how to execute all the algorithms that we introduce throughout the book. This will enable us to furnish an immediate quantitative assessment of their theoretical properties such as stability, accuracy and complexity. We will solve several problems that will be raised through exercises and examples, often stemming from specific applications.

---

[1] MATLAB is a trademark of TheMathWorks Inc., 24 Prime Park Way, Natick, MA 01760, Tel: 001+508-647-7000, Fax: 001+508-647-7001.

Several graphical devices will be adopted in order to render the reading more pleasant. We will report in the margin the MATLAB command along side the line where that command is being introduced for the first time. The symbol 🔧 will be used to indicate the presence of exercises, the symbol ⬦ to indicate the presence of a MATLAB program, while the symbol 💣 will be used when we want to attract the attention of the reader on a critical or surprising behavior of an algorithm or a procedure. The mathematical formulae of special relevance are put within a frame. Finally, the symbol 📖 indicates the presence of a display panel summarizing concepts and conclusions which have just been reported and drawn.

At the end of each chapter a specific section is devoted to mentioning those subjects which have not been addressed and indicate the bibliographical references for a more comprehensive treatment of the material that we have carried out.

Quite often we will refer to the textbook [QSS00] where many issues faced in this book are treated at a deeper level, and where theoretical results are proven. For a more thorough description of MATLAB we refer to [HH00]. All the programs introduced in this text can be downloaded from the web address

```
mox.polimi.it/Springer.
```

No special prerequisite is demanded of the reader, with the exception of an elementary course of Calculus.

However, in the course of the first chapter, we recall the principal results of Calculus and Geometry that will be used extensively throughout this text. The less elementary subjects, those which are not so necessary for an introductory educational path, are highlighted by the special symbol 🔍 .

We express our thanks to Thanh-Ha Le Thi from Springer-Verlag Heidelberg, and to Francesca Bonadei and Marina Forlizzi from Springer-Italia for their friendly collaboration throughout this project. We gratefully thank Prof. Eastham of Cardiff University for editing the language of the whole manuscript and stimulating us to clarify many points of our text.

Milano and Lausanne, May 2003   Alfio Quarteroni, Fausto Saleri

Contents

1. What can't be ignored

In this book we will systematically use elementary mathematical concepts which the reader should know already, yet he or she might not recall them immediately.

We will therefore use this chapter to refresh them, and to introduce new concepts as well which pertain to the field of Numerical Analysis. We will begin to explore their meaning and usefulness with the help of MATLAB (MATrix LABoratory), an integrated environment for programming and visualization in scientific computing. In Section 1.6 we will give a quick introduction to MATLAB, which is sufficient for the use that we are going to make in this book. However, we refer the interested readers to the manual [HH00] for a complete description of this language.

In the present Chapter we have therefore condensed notions which are typical of courses in Calculus, Linear Algebra and Geometry, yet rephrasing them in a way that is suitable for use in scientific computing.

1.1 Real numbers

While the set $\mathbb{R}$ of real numbers is known to everyone, the way in which computers treat them is perhaps less well known. On one hand, since machines have limited resources, only a subset $\mathbb{F}$ of finite dimension of $\mathbb{R}$ can be represented. These are called *floating-point numbers*. On the other hand, as we shall see in Section 1.1.2, $\mathbb{F}$ is characterized by properties that are different from those of $\mathbb{R}$. The reason is that any real number x is in principle truncated by the machine, giving rise to a new number (called the *floating point number*), denoted by $fl(x)$, which does not necessarily coincide with the original number x.

1.1.1 How do we represent them

To become acquainted with the differences between $\mathbb{R}$ and $\mathbb{F}$, let us make a few experiments using MATLAB which illustrate the way that a computer (e.g. a PC) deals with real numbers. Whether we use MATLAB rather than another language is just a matter of convenience. The results of our calculation, indeed, depend primarily on the manner in which the computer works, and only to a lesser degree on the programming language. Let us consider the rational number $x = 1/7$, whose decimal representation is $0.\overline{142857}$. This is an infinite representation, since the number of decimal digits is infinite. To get its computer representation, let us introduce after the *prompt* the ratio **1/7** and obtain

$>>$ 1/7
ans $=$
      0.1429

which is a number with only four decimal digits, the last being different from the fourth digit of the original number. Should we now consider 1/3 we would find 0.3333, so the fourth decimal digit would now be exact. This behavior is due to the fact that real numbers are *rounded* on the computer. This means, first of all, that only a fixed number of decimal digits are returned, and moreover the last decimal digit which appears is increased by unity whenever the first disregarded decimal digit is greater than or equal to 5.

The first remark to make is that using only four decimal digits to represent real numbers it is questionable. Indeed, the internal representation of the number is made with as many as 16 decimal digits, and what we have seen is simply one of several possible MATLAB output formats. The same number can take different expressions depending upon the specific **format** format declaration that is made. For instance, for the number 1/7, some possible output *formats* are:

format long	yields	0.14285714285714,
format short e	"	1.4286e − 01,
format long e	"	1.428571428571428e − 01,
format short g	"	0.14286,
format long g	"	0.142857142857143.

Some of them are more coherent than others with the internal computer representation. As a matter of fact, in general a computer stores a real number in the following way

$$x = (-1)^s \cdot (0.a_1 a_2 \ldots a_t) \cdot \beta^e = (-1)^s \cdot m \cdot \beta^{e-t}, \quad a_1 \neq 0, \qquad (1.1)$$

where s is either 0 or 1, β (a positive integer larger than or equal to 2) is the *basis* adopted by the specific computer at hand, m is an integer called the *mantissa* whose length t is the maximum number of digits a_i (ranging between 0 and $\beta-1$) that are stored, and e is an integral number called the *exponent*. The format `long e` is the one which most resembles this representation, and `e` stands for exponent; its digits, preceded by the sign, are reported to the right of the character `e`. The numbers whose form is given in (1.1) are called floating-point numbers, since the position of the decimal point is not fixed. The digits $a_1 a_2 \ldots a_p$ (with $p \le t$) are often called the first p significant digits of x.

The condition $a_1 \ne 0$ ensures that a number cannot have multiple representations. For instance, without this restriction the number $1/10$ could be represented (in the decimal basis) as $0.1 \cdot 10^0$, but also as $0.01 \cdot 10^1$, etc..

The set $\mathbb{F}$ is therefore completely characterized by the basis β, the number of significant digits t and the range (L, U) (with $L < 0$ and $U > 0$) of variation of the index e. Thus it is denoted as $\mathbb{F}(\beta, t, L, U)$. For instance, in MATLAB we have $\mathbb{F} = \mathbb{F}(2, 53, -1021, 1024)$ (indeed, 53 significant digits in basis 2 correspond to the 15 significant digits that are shown from MATLAB in basis 10 with the `format long`).

Fortunately, the *roundoff error* that is inevitably generated whenever a real number $x \ne 0$ is replaced by its representative $fl(x)$ in $\mathbb{F}$, is small, since

$$\frac{|x - fl(x)|}{|x|} \le \frac{1}{2}\epsilon_M, \qquad (1.2)$$

where $\epsilon_M = \beta^{1-t}$ provides the distance between 1 and its closest floating-point number different from 1. Note that ϵ_M depends on β and t. For instance, in MATLAB ϵ_M can be obtained through the command eps, **eps** and we obtain $\epsilon_M = 2^{-52} \simeq 2.22 \cdot 10^{-16}$. Let us point out that in (1.2) we estimate the *relative error* on x, which is undoubtedly more meaningful than the *absolute error* $|x - fl(x)|$. As a matter of fact, the latter doesn't account for the order of magnitude of x whereas the former does.

Number 0 does not belong to $\mathbb{F}$, as in that case we would have $a_1 = 0$ in (1.1): it is therefore handled separately. Moreover, L and U being finite, one cannot represent numbers whose absolute value is either arbitrarily large or arbitrarily small. Precisely, the smallest and the largest positive real numbers of $\mathbb{F}$ are given respectively by

$$x_{min} = \beta^{L-1}, \quad x_{max} = \beta^U (1 - \beta^{-t}).$$

realmin   In MATLAB these values can be obtained through the commands
realmax   realmin and realmax, yielding

$$x_{min} = 2.225073858507201 \cdot 10^{-308}$$
$$x_{max} = 1.7976931348623158 \cdot 10^{+308}.$$

A positive number smaller than x_{min} produces a message of under-flow and is treated either as 0 or in a special way (see, *e.g.*, [QSS00], Chapter 2). A positive number greater than x_{max} yields instead a mes-
Inf   sage of overflow and is stored in the variable Inf (which is the computer representation of $+\infty$).

The elements in $\mathbb{F}$ are more dense near x_{min}, and less dense while approaching x_{max}. As a matter of fact, the number in $\mathbb{F}$ nearest to x_{max} (to its left) and the one nearest to x_{min} (to its right) are, respectively

$$x_{max}^{-} = 1.7976931348623157 \cdot 10^{+308}$$
$$x_{min}^{+} = 2.225073858507202 \cdot 10^{-308}.$$

Thus $x_{min}^{+} - x_{min} \simeq 10^{-323}$, while $x_{max} - x_{max}^{-} \simeq 10^{292}$ (!). However, the relative distance is small in both cases, as we can infer from (1.2).

1.1.2   How do we operate with floating-point numbers

Since $\mathbb{F}$ is a proper subset of $\mathbb{R}$, elementary algebraic operations on float-ing point numbers do not enjoy all the properties of analogous oper-ations on $\mathbb{R}$. Precisely, commutativity still holds for addition (that is $fl(x + y) = fl(y + x)$) as well as for multiplication ($fl(xy) = fl(yx)$), but other properties such as associativity and distributivity are violated. Moreover, 0 is no longer unique. Indeed, let us assign the variable a the value 1, and execute the following instructions:

```
>> a = 1; b=1; while a+b ~= a; b=b/2; end
```

The variable b is halved at every step as long as the sum of a and b remains different (~=) from a. Should we operate on real numbers, this program would never end, whereas in our case it ends after a finite number of steps and returns the following value for b: 1.1102e-16= $\epsilon_M/2$. There exists therefore at least one number b different from 0 such that a+b=a. This is possible since $\mathbb{F}$ is made up of isolated numbers; when adding two numbers a and b with b<a and b less than ϵ_M, we always obtain that a+b is equal to a.

Associativity is violated whenever a situation of overflow or underflow occurs. Take for instance a=1.0e+308, b=1.1e+308 and c=-1.001e+308, and carry out the sum in two different ways. We find that

$$a + (b + c) = 1.0990e + 308, \quad (a + b) + c = \text{Inf}.$$

This is a particular instance of what occurs when one adds two numbers with opposite sign but similar absolute value. In this case the result may be quite inexact and the situation is referred to as *loss*, or *cancellation, of significant digits*. For instance, let us compute $((1 + x) - 1)/x$ (the obvious result being 1 for any $x \neq 0$):

```
>> x = 1.e - 15; ((1 + x) - 1)/x
ans =
       1.1102
```

This result is rather imprecise, the relative error being larger than 11%!.

Another case of numerical cancellation is encountered while evaluating the function

$$f(x) = x^7 - 7x^6 + 21x^5 - 35x^4 + 35x^3 - 21x^2 + 7x - 1 \qquad (1.3)$$

at 401 equispaced points with abscissa in $[1 - 2 \cdot 10^{-8}, 1 + 2 \cdot 10^{-8}]$. We obtain the chaotic graph reported in Figure 1.1 (the real behavior is that of $(x - 1)^7$, which is substantially constant and equal to the null function in such a tiny neighborhood of $x = 1$). In Section 1.4 we will see the commands that have generated the graph.

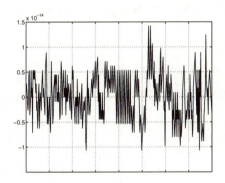

Fig. 1.1. Oscillatory behavior of the function (1.3) caused by cancellation errors

Finally, it is interesting to notice that in $\mathbb{F}$ there is no place for indeterminate forms such as $0/0$ or ∞/∞, whose presence produces what is called *not a number* (NaN in MATLAB) for which the normal rules of calculus do not apply. NaN

Remark 1.1 Whereas it is true that roundoff errors are usually small, when repeated within long and complex algorithms, they may give rise to catastrophic effects. Two outstanding cases concern the explosion of the Arianne missile on June 4, 1996, engendered by an overflow in the computer on board, and the falling on an American barrack of an American Patriot missile, during the Gulf War in 1991, because of a roundoff error in the computation of its trajectory.

An example with less catastrophic (but still troublesome) consequences is provided by the sequence

$$z_2 = 2, \quad z_{n+1} = 2^{n-1/2}\sqrt{1 - \sqrt{1 - 4^{1-n}z_n^2}}, \quad n = 2, 3, \ldots \quad (1.4)$$

which converges to π when n tends to infinity. When MATLAB is used to compute z_n, the relative error found between π and z_n decreases for the first 16 iterations , then grows because of roundoff errors (as shown in Figure 1.2).

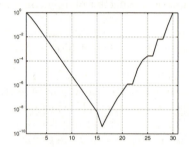

Fig. 1.2. Logarithm of the relative error $|\pi - z_n|/\pi$ versus n

See the Exercises 1.1-1.2.

1.2 Complex numbers

Complex numbers, whose set is denoted by $\mathbb{C}$, have the form $z = x + iy$ where $i = \sqrt{-1}$ is the imaginary unit (that is $i^2 = -1$), while $x = \text{Re}(z)$ and $y = \text{Im}(z)$ are the real and imaginary part of z, respectively. They are generally represented on the computer as pairs of real numbers.

Unless redefined otherwise, MATLAB variables i as well as j denote the imaginary unit. To introduce a complex number of real part x and imaginary part y, one can just write x+i*y; as an alternative, one can complex use the command complex(x,y). Let us also mention the trigonometric (or polar) representation of a complex number z,

$$z = \rho e^{i\theta} = \rho(\cos\theta + i\sin\theta), \quad (1.5)$$

where $\rho = \sqrt{x^2 + y^2}$ is the absolute value of the complex number (it can abs be obtained by setting abs(z)) and θ the argument, that is the angle between the vector z of components (x, y), and the x-axis. θ can be found angle by typing angle(z). The representation (1.5) is therefore:

$$\texttt{abs(z)} * (\cos(\texttt{angle(z)}) + \texttt{i} * \sin(\texttt{angle(z)})).$$

The polar representation of one or more complex numbers can be obtained through the command `compass(z)` where `z` is either a single   compass complex number or a vector whose components are complex numbers. For instance, by typing

```
>> z = 3+i*3; compass(z);
```

one obtains the graph reported in Figure 1.3.

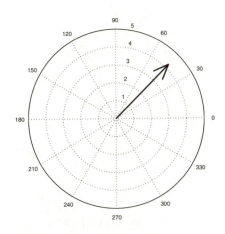

Fig. 1.3. Output of the MATLAB command `compass`

For any given complex number `z`, one can extract its real part with the command `real(z)` and its imaginary part with `imag(z)`. Finally, the   real complex conjugate $\bar{z} = x - iy$ of z, can be obtained by simply writing   imag `conj(z)`.   conj

In MATLAB *all operations are carried out by implicitly assuming that the operands as well as the result are complex.* We may therefore find some apparently surprising results. For instance, if we compute the cube root of -5 with the MATLAB command $(\texttt{-5})\texttt{\^{}(1/3)}$, instead of $-1.7099\ldots$ we obtain the complex number $0.8550 + 1.4809i$. (We anticipate the use of the symbol `^` for the power exponent.) As a matter of fact, all num-   ^ bers of the form $\rho e^{i(\theta+2k\pi)}$, with k an integer, are indistinguishable from z. By computing $\sqrt[3]{z}$ we find $\sqrt[3]{\rho}e^{i(\theta/3+2k\pi/3)}$, that is, the three distinct roots

$$z_1 = \sqrt[3]{\rho}e^{i\theta/3}, \quad z_2 = \sqrt[3]{\rho}e^{i(\theta/3+2\pi/3)}, \quad z_3 = \sqrt[3]{\rho}e^{i(\theta/3+4\pi/3)}.$$

MATLAB will select the one that is encountered by spanning the complex plane counterclockwise beginning from the real axis. Since the polar representation of $z = -5$ is $\rho e^{i\theta}$ with $\rho = 5$ and $\theta = -\pi$, the three roots are (see Fig. 1.4 for their representation in the Gauss plane)

$$z_1 = \sqrt[3]{5}(\cos(-\pi/3) + i\sin(-\pi/3)) \simeq 0.8550 - 1.4809i,$$

$$z_2 = \sqrt[3]{5}(\cos(\pi/3) + i\sin(\pi/3)) \simeq 0.8550 + 1.4809i,$$

$$z_3 = \sqrt[3]{5}(\cos(-\pi) + i\sin(-\pi)) \simeq -1.71.$$

The second root is the one which is selected.

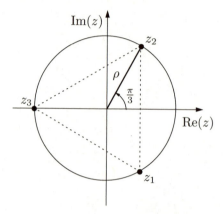

Fig. 1.4. Representation in the complex plane of the three cube roots of the real number -5

Finally, by (1.5) we obtain the Euler formula:

$$\cos(\theta) = \frac{1}{2}\left(e^{i\theta} + e^{-i\theta}\right), \quad \sin(\theta) = \frac{1}{2i}\left(e^{i\theta} - e^{-i\theta}\right). \tag{1.6}$$

1.3 Matrices

Let n and m be positive integers. A matrix with m rows and n columns is a set of $m \times n$ elements a_{ij}, with $i = 1, \ldots, m$, $j = 1, \ldots, n$, represented by the following table:

$$A = \begin{bmatrix} a_{11} & a_{12} & \cdots & a_{1n} \\ a_{21} & a_{22} & \cdots & a_{2n} \\ \vdots & \vdots & & \vdots \\ a_{m1} & a_{m2} & \cdots & a_{mn} \end{bmatrix}. \tag{1.7}$$

In compact form we write $A = (a_{ij})$. Should the elements of A be real numbers, we write $A \in \mathbb{R}^{m \times n}$, and $A \in \mathbb{C}^{m \times n}$ if they are complex.

Square matrices of dimension n are those with $m = n$. A matrix featuring a single column is a *column vector*, whereas a matrix featuring a single row is a *row vector*.

In order to introduce a matrix in MATLAB one has to write the elements from the first to the last row, introducing the character ; to separate the different rows. For instance, the command

```
>> A = [ 1 2 3; 4 5 6]
```

produces

```
A =
    1   2   3
    4   5   6
```

that is, a 2×3 matrix whose elements are indicated above. The null matrix 0 is that with null elements $a_{ij} = 0$ for $i = 1, \ldots, m$, $j = 1, \ldots, n$; it can be constructed by the MATLAB command `zeros(m,n)`. The MATLAB command `eye(m,n)` produces a rectangular matrix whose elements are all 0 except those on the main diagonal which are unity.

zeros
eye

The main diagonal of a $m \times n$ matrix A is the diagonal made of elements a_{ii}, $i = 1, \ldots, \min(m, n)$.

A particular case is the command `eye(n)` (which is a shorthand version for `eye(n,n)`); it produces a square matrix of dimension n which is called the *identity* matrix and is denoted by I.

We can define the following operations:

1. if $A = (a_{ij})$ and $B = (b_{ij})$ are $m \times n$ matrices, the *sum* of A and B is the matrix $A + B = (a_{ij} + b_{ij})$;

2. the *product* of a matrix A by a real or complex number λ is the matrix $\lambda A = (\lambda a_{ij})$;

3. the *product* of two matrices is possible only for compatible sizes, precisely if A is $m \times p$ and B is $p \times n$, for some positive integer p. In that case $C = AB$ is an $m \times n$ matrix whose elements are

$$c_{ij} = \sum_{k=1}^{p} a_{ik} b_{kj}, \quad \text{for } i = 1, \ldots, m, \; j = 1, \ldots, n.$$

Here is an example of the sum and product of two matrices.

```
>> A=[1 2 3; 4 5 6]; B=[7 8 9; 10 11 12]; C=[13 14; 15 16; 17 18];
>> A+B
ans =
    8   10   12
```

```
   14   16   18
>> A*C
ans =
   94  100
  229  244
```

Note that MATLAB returns a diagnostic message when one tries to carry out operations on matrices with incompatible dimensions. For instance:

```
>> A+C
??? Error using ==> +
Matrix dimensions must agree.
>> A*B
??? Error using ==> *
Inner matrix dimensions must agree.
```

size **Remark 1.2** For any given matrix A, the MATLAB command `dim=size(A)`
returns the two-element variable `dim=[m,n]` containing the number of rows
whos and columns of A. More in general, the command `whos` allows one to check the
type of all variables in use in a work session. For instance, if we have defined
only the matrices A, B and C as before, the command `whos` returns

Name	Size	Bytes	Class
A	2x3	48	double array
B	2x3	48	double array
C	3x2	48	double array

Grand total is 18 elements using 144 bytes

If A is a square matrix of dimension n, its *inverse* (provided it exists)
is a square matrix of dimension n, denoted by A^{-1}, which satisfies the
matrix relation $AA^{-1} = A^{-1}A = I$. We can obtain A^{-1} through the
inv command `inv(A)`. The inverse of A exist iff the *determinant* of A, a
number denoted by $\det(A)$, is non-zero. The latter condition is satisfied
in turn iff the column vectors of A are linearly independent (see Section
1.3.1). The determinant of a square matrix is defined by the following
recursion formula (*Laplace rule*):

$$\det(A) = \begin{cases} a_{11} & \text{if } n = 1, \\ \sum_{j=1}^{n} \Delta_{ij} a_{ij}, & \text{for } n > 1, \ \forall i = 1, \ldots, n, \end{cases} \tag{1.8}$$

where $\Delta_{ij} = (-1)^{i+j} \det(A_{ij})$ and A_{ij} is the matrix obtained by eliminating the i-th row and j-th column from matrix A. In particular, if

$A \in \mathbb{R}^{1\times 1}$ we set $\det(A) = a_{11}$; if $A \in \mathbb{R}^{2\times 2}$ one has

$$\det(A) = a_{11}a_{22} - a_{12}a_{21};$$

if $A \in \mathbb{R}^{3\times 3}$ we obtain

$$\det(A) = a_{11}a_{22}a_{33} + a_{31}a_{12}a_{23} + a_{21}a_{13}a_{32}$$
$$-a_{11}a_{23}a_{32} - a_{21}a_{12}a_{33} - a_{31}a_{13}a_{22}.$$

For the matrix product we have the following property: if $A = BC$, then $\det(A) = \det(B) \cdot \det(C)$.

To invert a 2×2 matrix and compute its determinant we can proceed as follows:

```
>> A=[1 2; 3 4];
>> inv(A)
ans =
  -2.0000    1.0000
   1.5000   -0.5000
>> det(A)
ans =
  -2
```

Should a matrix be singular, MATLAB returns a diagnostic message, followed by a matrix whose elements are all equal to `Inf`, as illustrated by the following example:

```
>> A=[1 2; 0 0];
>> inv(A)
Warning: Matrix is singular to working precision.
ans =
  Inf   Inf
  Inf   Inf
```

For special classes of square matrices, the computation of inverses and determinants is rather simple. In particular, if A is a *diagonal matrix*, i.e. one for which only the diagonal elements a_{kk}, $k = 1, \ldots, n$, are non-zero, its determinant is given by $\det(A) = a_{11}a_{22} \cdots a_{nn}$. In particular, A is non-singular iff $a_{kk} \neq 0$ for all k. In such a case the inverse of A is still a diagonal matrix with elements a_{kk}^{-1}.

Let v be a vector of dimension n. The MATLAB command `diag(v)`   diag produces a diagonal matrix whose elements are the components of vector v. The more general command `diag(v,m)` yields a square matrix of dimension `n+abs(m)` whose m-th upper diagonal (*i.e.* the diagonal made of elements of indices $i, i + m$) has elements equal to the components of v, while the remaining elements are null. Note that this extension is valid also when m is negative, in which case the only affected elements are those of lower diagonals.

For instance if v = [1 2 3] then:

```
>> A=diag(v,-1)
A =
     0   0   0   0
     1   0   0   0
     0   2   0   0
     0   0   3   0
```

Other special matrices are the *upper triangular* and *lower triangular* matrices. A square matrix of dimension n is *lower* (respectively, *upper*) *triangular* if all elements above (respectively, below) the main diagonal are zero. Its determinant is simply the product of the diagonal elements.

tril   Through the commands tril(A) and triu(A), one can extract from triu the matrix A of dimension n its lower and upper triangular part. Their extensions tril(A,m) or triu(A,m), with m ranging from -n and n, permit the extraction of the triangular part augmented by, or deprived of, m extradiagonals.

For instance, given the matrix A =[3 1 2; -1 3 4; -2 -1 3], by the command L1=tril(A) we obtain

```
L1 =
     3    0    0
    -1    3    0
    -2   -1    3
```

while, by L2=tril(A,-1), we obtain

```
L2 =
     0    0    0
    -1    0    0
    -2   -1    0
```

Finally, we recall that if A $\in \mathbb{R}^{m \times n}$ its transpose $A^T \in \mathbb{R}^{n \times m}$ is the matrix obtained by interchanging rows and columns of A. When $A = A^T$ the matrix A is called *symmetric*. The MATLAB notation for A' the transpose of A is A'.

1.3.1 Vectors

Vectors will be indicated in boldface; precisely, $\mathbf{v}$ will denote a column vector whose i-th component is denoted by v_i. When all components are real numbers we can write $\mathbf{v} \in \mathbb{R}^n$.

In MATLAB, vectors are regarded as particular cases of matrices. To introduce a column vector one has to insert between square brackets the values of its components separated by semi-colons, whereas for a row vector it suffices to write the component values separated by blanks or commas. For instance, through the instructions v = [1;2;3] and w = [1 2 3] we initialize the column vector $\mathbf{v}$ and the row vector $\mathbf{w}$, both of dimension 3. The command zeros(n,1) (respectively, zeros(1,n)) zeros produces a column (respectively, row) vector of dimension n with null elements, which we will denote by $\mathbf{0}$. Similarly, the command ones(n,1) ones generates the column vector, denoted with $\mathbf{1}$, whose components are all equal to 1. Finally, with the command v=[ ] we initialize an empty vector.

A system of vectors $\{\mathbf{y}_1, \ldots, \mathbf{y}_m\}$ is called *linearly independent* if the relation

$$\alpha_1 \mathbf{y}_1 + \ldots + \alpha_m \mathbf{y}_m = \mathbf{0}$$

implies that all coefficients $\alpha_1, \ldots, \alpha_m$ are null. A system $\mathcal{B} = \{\mathbf{y}_1, \ldots, \mathbf{y}_n\}$ of n linearly independent vectors in $\mathbb{R}^n$ (or $\mathbb{C}^n$) is a *basis* for $\mathbb{R}^n$ (or $\mathbb{C}^n$), that is, any vector $\mathbf{w}$ in $\mathbb{R}^n$ can be written as

$$\mathbf{w} = \sum_{k=1}^{n} w_k \mathbf{y}_k,$$

for a unique possible choice of the coefficients $\{w_k\}$. The latter are called the *components* of $\mathbf{w}$ with respect to the basis $\mathcal{B}$. For instance, the canonical basis of $\mathbb{R}^n$ is the set of vectors $\{\mathbf{e}_1, \ldots, \mathbf{e}_n\}$, where $\mathbf{e}_i$ has its i-th component equal to 1, and all other components equal to 0. Although not a unique basis of $\mathbb{R}^n$, the latter is the one which is normally used.

The *scalar product* of two vectors $\mathbf{v}, \mathbf{w} \in \mathbb{R}^n$ is defined as

$$(\mathbf{v}, \mathbf{w}) = \mathbf{w}^T \mathbf{v} = \sum_{k=1}^{n} v_k w_k,$$

$\{v_k\}$ and $\{w_k\}$ being the components of $\mathbf{v}$ and $\mathbf{w}$, respectively. The corresponding MATLAB command is w'*v, where now the apex denotes transposition of a vector. The length (or modulus) of a vector $\mathbf{v}$ is given

by

$$\|\mathbf{v}\| = \sqrt{(\mathbf{v}, \mathbf{v})} = \sqrt{\sum_{k=1}^{n} v_k^2}$$

norm   and can be computed through the command `norm(v)`.

.* The MATLAB command `x.*y` or `x.^2` indicates that these operations
.^ should be carried out component by component. For instance if we define
the vectors

>> v = [1; 2; 3]; w = [4; 5; 6];

the instruction

>> w'*v
ans =
    32

provides their scalar product, while

>> w.*v
ans =
    4   10   18

returns a vector whose i-th component is equal to $x_i y_i$.

Finally, we recall that a number λ (real or complex) is an *eigenvalue*
of the matrix $A \in \mathbb{R}^{n \times n}$, if

$$A\mathbf{v} = \lambda \mathbf{v},$$

for a suitable $\mathbf{v} \in \mathbb{C}^n$, $\mathbf{v} \neq \mathbf{0}$, which is called an *eigenvector* associated
with λ.

In general, the computation of eigenvalues is quite difficult. Exceptions
are represented by diagonal and triangular matrices, whose eigenvalues
are their diagonal elements.

See the Exercises 1.3-1.5.

1.4 Real functions

This chapter will deal with real functions. In particular, given a function
f which is defined on an interval (a, b), we would like to compute its
zeros, its integral and its derivative, as well as to determine its behavior.

fplot   The command `fplot(fun,lims)` plots the graph of the function `fun`
(which is stored as a string of characters) on the interval (`lims(1)`,`lims(2)`).
For instance, to represent $f(x) = 1/(1 + x^2)$ on $(-5, 5)$, we can write

```
>> fun ='1/(1+x^2)'; lims=[-5,5]; fplot(fun,lims);
```

or, more directly,

```
>> fplot('1/(1+x^2)',[-5 5]);
```

The graph is obtained by sampling the function on a set of non-equispaced abscissae and reproduces the true graph of f with a tolerance of 0.2%. To improve the accuracy we could use the command

```
>> fplot(fun,lims,tol,n,'LineSpec',P1,P2,...)
```

where `tol` indicates the desired tolerance and the parameter $n(\geq 1)$ ensures that the function will be plotted with a minimum of $n+1$ points. 'LineSpec' is a given line specification (for instance, '--' for a dashed line, ':' for a dotted line), while the parameters P1,P2,... can be passed directly to the function `fun`. To use default values for `tol`, n or 'LineSpec' one can pass empty matrices ([ ]).

To evaluate a function `fun` at a point x we write y=eval(fun), after having initialized x. The corresponding value is stored in y. Note that x, and correspondingly y, can be a vector. When using this command, the restriction is that the argument of the function `fun` must be x. When the argument of `fun` has a different name (this is often the case when this argument is generated at the interior of a program) the command eval would be replaced by `feval` (see Remark 1.4).   **eval**

Finally, we point out that if we write **grid on** after the command   **grid**
fplot, we can obtain the background-grid as that in Figure 1.1.

Remark 1.3 In general circumstances, it might be useful to allow any number of arguments to a function. This is made possible by using the variable **varargin**, a cell array containing the optional arguments to the function.   **varargin**
varargin must be declared as the last input argument and collects all the inputs from that point onwards.

For instance, the program (see Section 1.6.2)

```
function L=mytril(varargin)
L=tril(varargin{:});
```

collects all the inputs starting into the variable **varargin**. **mytril** uses the comma-separated list syntax **varargin{:}** to pass the parameters to the function **tril**. The call,

```
L=mytril([1 2 3; 4 5 6; 7 8 9],-2)
```

```
L =
    0    0    0
    0    0    0
    7    0    0
```

results in `varargin` being a 1-by-2 cell array containing the values [1 2 3; 3 4 5; 6 7 8] and -2.

1.4.1   The zeros

We recall that α is said to be a zero of a real function f if $f(\alpha) = 0$. It is *simple* if $f'(\alpha) \neq 0$, *multiple* otherwise.

From the graph of a function one can infer (within a certain tolerance) which are its real zeros. The direct computation of all zeros of a given function is not always straightforward. For functions which are polynomials with real coefficients of degree n, that is, of the form

$$p_n(x) = a_0 + a_1 x + a_2 x^2 + \ldots + a_n x^n = \sum_{k=0}^{n} a_k x^k, \quad a_k \in \mathbb{R}, \ a_n \neq 0,$$

we can obtain the only zero $\alpha = -a_0/a_1$, when $n = 1$ (i.e. p_1 represents a straight line), or the two zeros, α_+ and α_-, when $n = 2$ (this time p_2 represents a parabola)

$$\alpha_{\pm} = \frac{-a_1 \pm \sqrt{a_1^2 - 4a_0 a_2}}{2a_2}.$$

However, there are no explicit formulae for the zeros of an arbitrary polynomial p_n when $n \geq 5$.

Also the number of zeros of a function cannot in general be determined *a priori*. An exception is provided by polynomials, for which the number of zeros (real or complex) coincides with the polynomial degree. Moreover, should $\alpha = x + iy$ be a zero of a polynomial, its complex conjugate $\bar{\alpha} = x - iy$ is also a zero.

To compute in MATLAB one zero of a function `fun`, near a given
`fzero` value `x0`, either real or complex, the command `fzero(fun,x0)` can be used. The result is an approximate value of the desired zero, and also the interval in which the search was made. Alternatively, using the command `fzero(fun,[x0 x1])`, a zero of `fun` is searched for in the interval whose extremes are `x0,x1`, provided f changes sign between `x0` and `x1`.

Let us consider, for instance, the function $f(x) = x^2 - 1 + e^x$. Looking at its graph we see that there are two zeros in $(-1, 1)$. To compute them we need to execute the following commands:

```
>> fun='x^2 - 1 + exp(x)';
>> fzero(fun,1)
Zero found in the interval: [-0.28, 1.9051].
ans =
```

  6.0953e-18
>> fzero(fun,-1)

Zero found in the interval: [-1.2263, -0.68].

ans =
  -0.7146

  In Chapter 2 we will introduce and investigate several methods for the approximate computation of the zeros of an arbitrary function.

1.4.2   Polynomials

Polynomials are very special functions and there is a special MATLAB toolbox, polyfun, for their treatment. The command polyval, is apt to   polyval
evaluate a polynomial at one or several points. Its input arguments are a vector p and a vector x, where the components of p are the polynomial coefficients stored in decreasing order, from a_n down to a_0, and the components of x are the abscissae where the polynomial needs to be evaluated. The result can be stored in a vector y by writing

>> y = polyval(p,x)

For instance, the values of $p(x) = x^7 + 3x^2 - 1$, at the equispaced abscissae $x_k = -1 + k * 0.25$ for $k = 0, \ldots, 8$, can be obtained by proceeding as follows:

>> p = [1 0 0 0 0 3 0 -1]; x = [-1:0.25:1];
>> y = polyval(p,x)
y =
  Columns 1 through 7
   1.0000   0.5540   -0.2578   -0.8126   -1.0000   -0.8124   -0.2422
  Columns 8 through 9
   0.8210   3.0000

  Alternatively, one could use the command fplot. However, in such case one should provide the entire analytic expression of the polynomial in the input string, and not simply its coefficients.
  Let us recall that if α is such that $p(\alpha) = 0$, then α is called zero of p or, equivalently, a *root* of the algebraic equation $p(x) = 0$. The program roots provides an approximation of the zeros of a polynomial   roots
and requires only the input of the vector p.
  For instance, we can compute the zeros of $p(x) = x^3 - 6x^2 + 11x - 6$ by writing

```
>> p = [1 -6 11 -6]; format long;
>> roots(p)
ans =
    3.00000000000000
    2.00000000000000
    1.00000000000000
```

Unfortunately, the result is not always that accurate. For instance, for the polynomial $p(x) = (x - 1)^7$, whose unique zero is $\alpha = 1$, we find (quite surprisingly)

```
>> p = [1 -7  21 -35  35  -21  7  -1];
>> roots(p)
ans =
    1.0088
    1.0055 + 0.0069i
    1.0055 - 0.0069i
    0.9980 + 0.0085i
    0.9980 - 0.0085i
    0.9921 + 0.0038i
    0.9921 - 0.0038i
```

This inaccuracy is due to the fact that the coefficients of p have alternating signs, yielding severe cancellation errors.

conv   We mention that using the command p=conv(p1,p2) we can obtain the coefficients of the polynomial given by the product of two polynomials whose coefficients are contained in the vectors p1 and p2. Similarly, the command [q,r]=deconv(p1,p2) provides the coefficients of the polynomials obtained on dividing p1 by p2, *i.e.* p1 = conv(p2,q) + r. In other words, q and r are the quotient and the reminder of the division.

deconv

Let us consider for instance the product and the ratio between the two polynomials $p_1(x) = x^4 - 1$ and $p_2(x) = x^3 - 1$:

```
>> p1 = [1 0 0 0 -1];
>> p2 = [1 0 0 -1];
>> p=conv(p1,p2)
p =
    1    0    0   -1   -1    0    0    1
>> [q,r]=deconv(p1,p2)
q =
    1    0
r =
    0    0    0    1   -1
```

We therefore find the polynomials $p(x) = p_1(x)p_2(x) = x^7 - x^4 - x^3 + 1$, $q(x) = x$ and $r(x) = x - 1$ such that $p_1(x) = q(x)p_2(x) + r(x)$.

Finally, the commands `polyint(p)` and `polyder(p)` provides respec- `polyint`
tively the coefficients of the primitive (vanishing at $x = 0$) and those `polyder`
of the derivative of the polynomial whose coefficients are given by the
components of the vector `p`.

If `x` is a vector of abscissae and `p` (respectively, p_i) is a vector containing the coefficients of a polynomial P (respectively, P_i), the previous
commands are summarized in the following table:

command	yields
`y=polyval(p,x)`	$y = $ values of $P(x)$
`z=roots(p)`	$z = $ roots of P such that $P(z) = 0$
`p=conv(p1,p2)`	$p = $ coefficients of the polynomial $P_1 P_2$
`[q,r]=deconv(p1,p2)`	$q = $ coefficients of Q, $r = $ coefficients of R such that $P_1 = QP_2 + R$
`y=polyder(p)`	$y = $ coefficients of $P'(x)$
`y=polyint(p)`	$y = $ coefficients of $\int P(x)\, dx$

A further command, `polyfit`, allows the computation of the $n+1$ polynomial coefficients of a polynomial P of degree n once the values attained
by P at $n+1$ distinct nodes are available (see Section 3.1.1).

1.4.3   Integration and differentiation

The following two results will often be invoked throughout this book.

1. the *fundamental theorem of integration*: if f is a continuous function in $[a, b)$, then

$$F(x) = \int_a^x f(t)\, dt$$

is a differentiable function, called a *primitive* of f, which satisfies,
$\forall x \in [a, b)$,

$$F'(x) = f(x);$$

2. the *first mean-value theorem for integrals*: if f is a continuous function in $[a, b)$ and $x_1, x_2 \in [a, b)$, then $\exists \xi \in (x_1, x_2)$ such that

$$f(\xi) = \frac{1}{x_2 - x_1} \int_{x_1}^{x_2} f(t)\, dt.$$

Even when it does exist, a primitive might be either impossible to determine or difficult to compute. For instance, knowing that $\ln|x|$ is a primitive of $1/x$ is irrelevant if one doesn't know how to compute efficiently the logarithms. In chapter 4 we will introduce several methods to compute the integral of arbitrary continuous functions with a desired accuracy, irrespectively of the knowledge of its primitive.

We recall that a function f defined on an interval $[a, b]$ is differentiable in a point $\bar{x} \in (a, b)$ if the following limit exists and is finite

$$f'(\bar{x}) = \lim_{h \to 0} \frac{1}{h}(f(\bar{x} + h) - f(\bar{x})). \tag{1.9}$$

If $\bar{x} = a$ this definition requires h to be positive, whereas h must be negative if $\bar{x} = b$. In all cases, the value of the derivative provides the slope of the tangent line to the graph of f at the point $\bar{x}$. We say that a function which is continuous together with its derivative at any point of $[a, b]$ belongs to the space $C^1([a, b])$. More generally, a function with continuous derivatives up to the order p (a positive integer) is said to belong to $C^p([a, b])$. In particular, $C^0([a, b])$ denotes the space of continuous functions in $[a, b]$.

A result that will be often used is the *mean value theorem*, according to which, if $f \in C^1([a, b])$, there exists $\xi \in (a, b)$ such that

$$f'(\xi) = (f(b) - f(a))/(b - a).$$

Finally, it is worth recalling that a function continuous with all its derivatives up to the order $n + 1$ in a neighborhood of x_0, can be approximated in such a neighborhood by the so-called *Taylor polynomial of degree n at the point x_0*:

$$\begin{aligned} T_n(x) &= f(x_0) + (x - x_0)f'(x_0) + \ldots + \frac{1}{n!}(x - x_0)^n f^{(n)}(x_0) \\ &= \sum_{k=0}^{n} \frac{(x - x_0)^k}{k!} f^{(k)}(x_0). \end{aligned}$$

diff   The MATLAB toolbox `symbolic` provides the commands `diff`, `int`
int   and `taylor` which allow us to obtain the analytical expression of the
taylor   derivative, the indefinite integral (i.e. a primitive) and the Taylor polynomial, respectively, of a given function. In particular, having defined in the string `f` the function on which we intend to operate, `diff(f,n)` provides its derivative of order `n`, `int(f)` its indefinite integral, and `taylor(f,x,n+1)` the associated Taylor polynomial of degree `n` in a neighborhood of $x_0 = 0$. The variable `x` must be declared *symbolic* by
syms   using the command `syms x`. This will allow its algebraic manipulation

without specifying its value.

In order to do this for the function $f(x) = (x^2 + 2x + 2)/(x^2 - 1)$, we proceed as follows:

```
>> f = '(x^2+2*x+2)/(x^2-1)';
>> syms x
>> diff(f)
(2*x+2)/(x^2-1)-2*(x^2+2*x+2)/(x^2-1)^2*x
>> int(f)
x+5/2*log(x-1)-1/2*log(1+x)
>> taylor(f,x,6)
-2-2*x-3*x^2-2*x^3-3*x^4-2*x^5
```

The command `funtool`, by the graphical interface illustrated in Fig. `funtool` 1.5, allows a very easy symbolic manipulation of arbitrary functions.

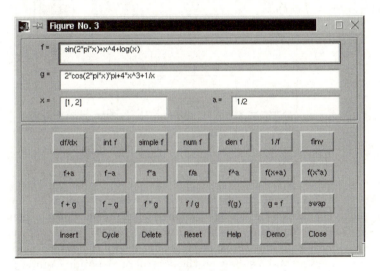

Fig. 1.5. Graphical interface of the command `funtool`

See the Exercises 1.6-1.7.

1.5   To err is not only human

As a matter of fact, by re-phrasing the Latin motto *Errare humanum est*, we might say that in numerical computation to err is even inevitable.

As we have seen, the simple fact of using a computer to represent real numbers introduces errors. What is therefore important is not to strive to eliminate errors, but rather to be able to control their effect.

Generally speaking, we can identify several levels of errors that occur during the approximation and resolution of a physical problem (see Figure 1.6).

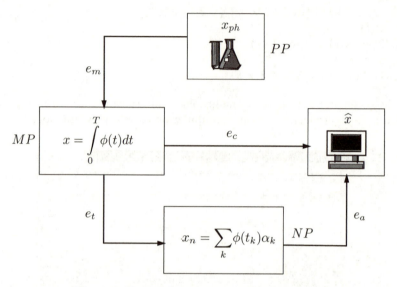

Fig. 1.6. Types of errors in a computational process

At the highest level stands the error e_m which occurs when forcing the physical reality (PP stands for physical problem and x_{ph} denotes its solution) to obey some mathematical model (MP, whose solution is x). Such errors will limit the applicability of the mathematical model to certain situations and are beyond the control of Scientific Computing.

The mathematical model (whether expressed by an integral as in the example of Figure 1.6, an algebraic or differential equation, a linear or non linear system) is generally not solvable in explicit form. Its resolution by computer algorithms will surely involve the introduction and propagation of roundoff errors at least. Let's call these errors e_a.

On the other hand, it is often necessary to introduce further errors since any procedure of the mathematical model involving an infinite sequence of arithmetic operations cannot be performed by the computer unless approximately. For instance the computation of the sum of a series will necessarily be accomplished in an approximate way by considering a suitable truncation.

It will therefore be necessary to introduce a numerical problem , NP, whose solution x_n differs from x by an error e_t which is called *truncation error*. Such errors do not occur only in mathematical models that are already set in finite dimension (for instance, when solving a linear system). The sum of the errors e_a and e_t constitutes the *computational error* e_c, the quantity we are interested in.

The *absolute* computational error is the difference between x, the exact solution of the mathematical model, and $\widehat{x}$, the solution obtained at the end of the numerical process,

$$e_c^{abs} = |x - \widehat{x}|,$$

while (if $x \neq 0$) the *relative* computational error is

$$e_c^{rel} = |x - \widehat{x}|/|x|,$$

where $|\cdot|$ denotes the modulus, or other measure of size, depending on the meaning of x.

The numerical process is generally an approximation of the mathematical model obtained as a function of a discretization parameter, which we will refer to as h and suppose positive. If, as h tends to 0, the numerical process returns the solution of the mathematical model, we will say that the numerical process is *convergent*. Moreover, if the (absolute or relative) error can be expressed as a function of h as

$$e_c = Ch^p, \tag{1.10}$$

where C is independent of h and p is a positive number, we will say that the method is *convergent of order p*.

Example 1.1 Suppose we approximate the derivative of a function f at a point $\bar{x}$ with the incremental ratio that appears in (1.9). Obviously, if f is differentiable at $\bar{x}$, the error committed by replacing f' by the incremental ratio tends to 0 as $h \to 0$. However, as we will see in Section 4.1, the error can be considered as Ch only if $f \in C^2$ in a neighborhood of $\bar{x}$.

While studying the convergence properties of a numerical procedure we will often deal with graphs reporting the error as a function of h in a logarithmic scale, which shows $\log(h)$ on the abscissae axis and $\log(e_c)$ on the ordinates axis. The purpose of this representation is easy to see: if $e_c = Ch^p$ then $\log e_c = \log C + p \log h$. p in logarithmic scale therefore represents the slope of the straight line $\log e_c$, so if we must compare two methods, the one presenting the greater slope will be the one with a higher order. In MATLAB it is very simple to obtain graphs in a logarithmic scale: one just needs to type `loglog(x,y)`, x and y being `loglog` the vectors containing the abscissae and the ordinates of the data to be represented.

For instance, in Figure 1.7 are reported the straight lines relative to the behavior of the errors in two different methods. The continuous line represents a first-order approximation, while the hatched line represents a second-order one.

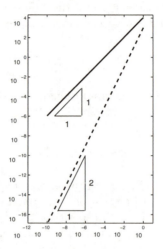

Fig. 1.7. Plot in logarithmic scales

There is an alternative to the graphical way of establishing the order of a method when one knows the errors e_i relative to some given values h_i of the parameter of discretization, with $i = 1, \ldots, N$: it consists in supposing that e_i is equal to Ch_i^p where C does not depend on i. One can then approach p with the values:

$$p_i = \log(e_i/e_{i-1})/\log(h_i/h_{i-1}) \quad i = 2, \ldots, N. \tag{1.11}$$

Actually the error is not a computable quantity since it depends on the unknown solution. Therefore it is necessary to introduce computable quantities that can be used to estimate the error itself, the so called *error estimator*. We will see some examples in Sections 2.2, 2.3 and 4.3.

1.5.1 Talking about costs

In general a problem is solved on the computer by an algorithm, which is a precise directive in the form of a finite text specifying the execution of a finite series of elementary operations. We are interested in those algorithms which involve only a finite number of steps.

The *computational cost* of an algorithm is the number of floating-point operations (in short, *ops*) that are required for its execution. Often, the speed of a computer is measured by the maximum number of floating-point operations which the computer can execute in one second (*flops*). In particular, the following abridged notations are commonly used: Mega-flops, equal to 10^6 *flops*, Giga-flops equal to 10^9 *flops*, Tera-flops equal to 10^{12} *flops*. The fastest computers nowadays reach as many as 40 of Tera-flops. Earlier versions of MATLAB allowed the count of the number of floating point operations by means of the command `flops`. This is no longer possible in MATLAB 6. However, through this book we will sometimes make use of the command `flops`; in those cases it is understood that the version 5.3 of MATLAB is being used.

In general, the exact knowledge of the number of operations requested by a given algorithm is not essential. Rather, it is useful to determine its order of magnitude as a function of a parameter d which is related to the problem dimension. We therefore say that an algorithm has *constant* complexity if it requires a number of operations independent of d, *i.e.* $\mathcal{O}(1)$ operations, *linear* complexity if it requires $\mathcal{O}(d)$ operations, or, more generally, *polynomial* complexity if it requires $\mathcal{O}(d^m)$ operations, for a positive integer m. Other algorithms may have *exponential* ($\mathcal{O}(c^d)$ operations) or even *factorial* ($\mathcal{O}(d!)$ operations) complexity. We recall that the symbol $\mathcal{O}(d^m)$ means "it behaves, for large d, like a constant times d^m".

Example 1.2 (matrix-vector product) Le A be a square matrix of order n and let $\mathbf{v} \in \mathbb{R}^n$: the $j - th$ component of the product $A\mathbf{v}$ is given by

$$a_{j1}v_1 + a_{j2}v_2 + \ldots + a_{jn}v_n,$$

and requires n products and $n - 1$ additions. One needs therefore $n(2n - 1)$ *ops* to compute all the components. Thus this algorithm requires $\mathcal{O}(n^2)$ *ops*, so it has a quadratic complexity with respect to the parameter n. The same algorithm would require $\mathcal{O}(n^3)$ *ops* to compute the product of two matrices of order n. However, there is an algorithm, due to Strassen, which requires "only" $\mathcal{O}(n^{\log_2 7})$ *ops*, and another, due to Winograd and Coppersmith, requiring $\mathcal{O}(n^{2.376})$ *ops*.

Example 1.3 (computation of a matrix determinant) As already mentioned, the determinant of a square matrix of order n can be computed by the recursive formula (1.8). The corresponding algorithm has a factorial complexity with respect to n and would be usable only for matrices of small dimension. For instance, if $n = 24$, a computer capable of performing as many as 1 Peta-flops (i.e. 10^{15} *ops* per second) would require 20 years to carry out this computation. One has therefore to resort to more efficient algorithms. Indeed, there exists an algorithm that allows the computation of determinants through matrix-matrix products, with henceforth a complexity of $\mathcal{O}(n^{\log_2 7})$ *ops* by resorting to the Strassen algorithm previously mentioned (see [BB96]).

The number of operations is not the sole parameter which matters in the analysis of an algorithm. Another relevant factor is represented by the time that is needed to access the computer memory (which depends on the way the algorithm has been coded). An indicator of the performance of an algorithm is therefore the CPU time (CPU stands for *central processing unit*), and can be obtained using the MATLAB command `cputime`. The total elapsed time between the *input* and *output* phases can be obtained by the command `etime`.

cputime

etime

Example 1.4 In order to compute the time needed for a matrix-vector multiplication we set up the following program:

```
>> A = rand(n,n); v = rand(n); T = [ ]; sizeA = [ ]; count = 1;
>> for k = 1:step:n
    AA = A(1:k,1:k); vv = v(1:k)';
    t = cputime;  b = AA*vv; tt = cputime - t;
    T = [T, tt]; sizeA = [sizeA,k]; count = count + 1;
end
```

a:step:b

rand

The instruction `a:step:b` appearing in the `for` cycle generates all numbers having the form `a+step*k` where k is an integer ranging from 0 to the largest value `kmax` for which `a+step*kmax` is not greater than b (in the case at hand, `a=1`, `b=2000` and `step=50`). The command `rand(n,m)` defines an n×m matrix of random entries. Finally, T is the vector whose components contain the time of CPU needed to carry out every single matrix-vector product, whereas `cputime` returns the CPU time in seconds that has been used by the MATLAB process since MATLAB started. The time necessary to execute a single program is therefore the difference between the actual CPU time and the one computed before the execution of the current program which is stored in the variable t. Figure 1.8, which is obtained by the command `plot(sizeA,T,'o')`, shows that the CPU time grows like the square of the matrix order n.

1.6 A few more words about MATLAB

MATLAB is an integrated environment for scientific computing and visualization. It is written in C language and is distributed by The Math-Works (see the website `www.mathworks.com`).

The main program is contained in the subdirectory *bin* of the principal directory *matlab*. Once installed, the execution of MATLAB allows access to a working environment characterized by the *prompt* >>. For instance, by executing MATLAB on our personal computer it shows

>>

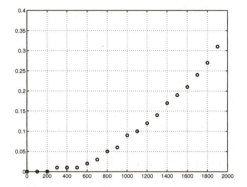

Fig. 1.8. Matrix-vector product: the CPU time (in seconds) versus the dimension n of the matrix (on a PC at 433 Mhz)

< M A T L A B >
Copyright 1984-2001 The MathWorks, Inc.
Version 6.1.0.450 Release 12.1
May 18 2001

To get started, select "MATLAB Help" from the Help menu.

>>

After pressing the key *enter* (or else *return*), all that is written after the *prompt* will be interpreted. [1] Precisely, MATLAB will first check whether what is written corresponds either to variables which have already been defined or to the name of one of the programs or commands defined in MATLAB. Should all those checks fail, MATLAB returns an error warning. Otherwise, the command is executed and an *output* will possibly be displayed. In all cases, at the end the system returns the *prompt* to acknowledge that it is ready for a new command. To close a MATLAB session one should write the command `quit` (or else `exit`) and press the key *enter*. From now it will be understood that to execute a program or a command one has to press the key *enter*. Moreover, the terms program, function or command will be used in an equivalent manner. When our command coincides with one of the elementary structures characterizing MATLAB (e.g. a number or a string of characters that are put among apices) they are immediately returned in *output* in the *default* variable `ans` (abbreviation of *answer*). Here is an example:

`quit`

`exit`

`ans`

---

[1]Thus a MATLAB program does not necessarily have to be compiled as other languages do, *e.g.* Fortran or C, although a MATLAB compiler may be invoked by the command mcc to allow a faster execution

```
>> 'home'
ans =
home
```

If now we write a different string (or number), **ans** will assume this new value.

We can turn off the automatic display of the *output* by writing a semicolon after the string. Thus if we write **'home'**; MATLAB will simply return the *prompt* (yet assigning the value **'home'** to the variable **ans**).

= More generally, the command = allows the assignment of a value (or a string of characters) to a given variable. For instance, to assign the string **'Welcome to NYC'** to the variable **a** we can write

```
>> a='Welcome to NYC';
```

As we can see, there is no need to declare the *type* of a variable, MATLAB will do it automatically and dynamically. For instance, should we write a=5, the variable **a** will now contain a number and no longer a string of characters. This flexibility is not cost-free. If we set a variable named **quit** equal to the number 5 we are inhibiting the use of the MATLAB command **quit**. We should therefore try to avoid using variables having

clear the name of MATLAB commands. However, by the command **clear** followed by the name of a variable (e.g. **quit**), it is possible to cancel this assignment and restore the original meaning of the command **quit**.

save Using the command **save** followed by the name **fname** all existing workspace variables are saved in the binary MATLAB file **fname.mat**.

load These data may be retrieved with the command **load fname.mat**. Omitting the filename causes **save** (or **load**) to use the default filename **matlab.mat**. To save the variables **v1, v2, ..., vn** the synthax is:

save fname v1 v2 ... vn.

help By the command **help** one can see the whole family of commands and pre-defined variables, including the so-called *toolboxes* which are sets of specialized commands. Among them let us recall those which define the

sin cos elementary functions such as sine (**sin(a)**), cosine (**cos(a)**), square root
sqrt exp (**sqrt(a)**), exponential (**exp(a)**).

There are special characters that cannot appear in the name of a
+ - * variable or in a command, for instance the algebraic operators (+, -,
/ & | ~ * and /), the logical operators *and* (&), *or* (|), *not* (~), the relational
> >= < operators *greater than* (>), *greater than or equal to* (>=), *less than* (<),
<= == *less than or equal to* (<=), *equal to* (==). Finally, a name can never begin with a digit, a bracket or with any punctuation mark.

1.6.1   MATLAB statements

A special programming language, the MATLAB language, is also available enabling the users to write new programs. Although its knowledge is not required for understanding how to use the several programs which we will introduce throughout this book, it may provide the reader with the capability of modifying them as well as producing new ones.

The MATLAB language features standard statements, such as conditionals and loops.

The *if-elseif-else* conditional has the following general form:

```
if cond(1)
    statement(1)
elseif cond(2)
    statement(2)
  .

  .

  .
else
    statement(n)
end
```

where cond(1), cond(2), ... represent MATLAB sets of instructions, with values 0 or 1 (false or true) and the entire construction allows the execution of that statement corresponding to the condition taking value equal to 1. Should all conditions be false, the execution of statement(n) will take place. This means that if the value of cond(k) is zero, the control moves on.

For instance, to compute the roots of a quadratic polynomial $ax^2 + bx + c$ one can use the following instructions (the command disp(.) simply displays what is written between brackets):

$$
\begin{aligned}
&>> \text{if a}\~= 0 \\
&\quad \text{sq} = \text{sqrt}(b*b - 4*a*c); \\
&\quad x(1) = 0.5*(-b + sq)/a; \\
&\quad x(2) = 0.5*(-b - sq)/a; \\
&\text{elseif b}\~= 0 \\
&\quad x(1) = -c/b; \\
&\text{elseif c}\~= 0 \\
&\quad \text{disp('Impossible equation');} \\
&\text{else} \\
&\quad \text{disp('The given equation is an identity')} \\
&\text{end}
\end{aligned}
\tag{1.12}
$$

Note that MATLAB does not execute the entire construction until the statement **end** is typed.

MATLAB allows two types of loops, a *for-loop* (comparable to a Fortran *do-loop* or a C *for-loop*) and a *while-loop*. A for-loop repeats the statements in the loop as the loop index takes on the values in a given row vector. For instance, to compute the first 6 terms of the Fibonacci sequence $\{f_i = f_{i-1} + f_{i-2}\}$ with $f_1 = 0$ and $f_2 = 1$, one can use the following instructions:

```
>> f(1) = 0; f(2) = 1;
>> for i = [3 4 5 6]
     f(i) = f(i-1) + f(i-2);
   end
```

Note that a semicolon can be used to separate several MATLAB instructions typed on the same line. Also, note that we can replace the second instruction by the equivalent **>> for i = 3:6**. The while-loop repeats as long as the given **cond** is true. For instance, the following set of instructions can be used as an alternative to the previous set:

```
>> f(1) = 0; f(2) = 1; k = 3;
>> while k <= 6
     f(k) = f(k-1) + f(k-2); k = k + 1;
   end
```

Other statements of perhaps less frequent use exist, such as *switch, case, otherwise*. The interested reader can have access to their meaning by the **help** command.

1.6.2 Programming in MATLAB

Let us now explain briefly how to write MATLAB programs. As previously noticed, new programs may be added to MATLAB. A new program must be put in a file with a given name with extension **m**, which is called *M-file*. They must be located in one of the directories in which MATLAB automatically searches for m-files; their list can be obtained by the command **path** (see **help path** to learn how to add a directory to this list). **path** The first directory scanned by MATLAB is the current working directory.

It is important at this level to distinguish between *scripts* and *functions*. A script is simply a collection of MATLAB commands in an *m-file* and can be used interactively. For instance, the set of instructions (1.12) can give rise to a script (which we could name **equation**) by copying it in the file **equation.m**. To launch it, one can simply write after the

MATLAB prompt >> the instruction `equation`. We report two examples below:

```
>> a = 1; b = 1; c = 1;
>> equation
ans =
 -0.5000 + 0.8660i  -0.5000 - 0.8660i

>> a = 0; b = 1; c = 1;
>> equation
ans =
   -1
```

Much more flexible than scripts are *functions*. A function is, *in general*, defined in an m-file (which we will generically call `name.m`) beginning with a line of the following form:

```
function [out1,...,outn]=name(in1,...,inm)
```

where `out1,...,outn` are the output variables and `in1,...,inm` are the input variables.

The following file, called `det23.m`, defines a new function called `det23` which computes, according to the formulae given in Section 1.3, the determinant of a matrix whose dimension could be either 2 or 3:

```
function det=det23(A)
%DET23 computes the determinant of a square matrix
% of dimension 2 or 3
[n,m]=size(A);
if n==m
  if n==2
    det = A(1,1)*A(2,2)-A(2,1)*A(1,2);
  elseif n == 3
    det = A(1,1)*det23(A[2,3],[2,3]))-A(1,2)*det23(A([2,3],[1,3]))+...
        A(1,3)*det23(A[2,3],[1,2]));
  else
    disp(' Only 2x2 or 3x3 matrices ');
  end
else
  disp(' Only square matrices ');
end
return
```

Note the use of the continuation characters ... meaning that the in- ... struction is continuing on the next line. The instruction `A([i,j],[k,l])` allows the construction of a 2×2 matrix whose elements are the elements of the original matrix A lying at the intersections of the i-th and j-th rows with the k-th and l-th columns.

When a function is invoked, MATLAB creates a local workspace. The commands in the function cannot refer to variables from the global (interactive) workspace unless they are passed as inputs. In particular, variables used in a function are erased when the execution terminates, unless they are returned as output parameters. The symbol % is used to begin comments.

Usually functions terminate when the end of the function is reached, **return** however a **return** statement can be used to force an early return (according to the fulfillment of a certain condition). For instance, in order to approximate the golden section number $\alpha = 1.6180339887\ldots$, which is the limit for $k \to \infty$ of the quotient f_k/f_{k-1}, by iterating until the difference between two consecutive ratios is less than 10^{-4}, we can construct the following function:

```
function [golden,k]=fibonacci
f(1) = 0; f(2) = 1; goldenold = 0; kmax = 100; tol = 1.e-04;
for k = 3:kmax
   f(k) = f(k-1) + f(k-2);
   golden = f(k)/f(k-1);
   if abs(golden - goldenold) <= tol
      return
   end
   goldenold = golden;
end
return
```

Then, we can write

```
>> [alpha,niter]=fibonacci
alpha =
   1.61805555555556
niter =
   14
```

After 14 iterations the function has returned an approximate value which shares with α the first 5 significant digits.

The number of input and output parameters of a MATLAB function can vary. For instance, we could modify the Fibonacci function as follows:

```
function [golden,k]=fibonacci(tol,kmax)
if nargin == 0
   kmax = 100; tol = 1.e-04; % default values
elseif nargin == 1
```

```
    kmax = 100; % default value only for kmax
end
f(1) = 0; f(2) = 1; goldenold = 0;
for k = 3:kmax
    f(k) = f(k-1) + f(k-2);
    golden = f(k)/f(k-1);
    if abs(golden - goldenold) <= tol
        return
    end
    goldenold = golden;
end
return
```

The function **nargin** counts the number of input parameters. In the   nargin
new version of the function **fibonacci** we can prescribe the maximum
number of inner iterations allowed (**kmax**) and a specific tolerance **tol**.
When this information is missing the function must provide default values (in our case, kmax = 100 and tol = 1.e-04). A possible use of it is
as follows:

```
>> [alpha,niter]=fibonacci(1.e-6,200)
alpha =
    1.61803381340013
niter =
    19
```

Note that using a stricter tolerance we have obtained a new approximate
value that shares with α as many as 8 significant digits.
The function **nargin** can be used externally to a given function to obtain
the number of input parameters. Here is an example:

```
>> nargin('fibonacci')
ans =
    2
```

Remark 1.4 (inline functions) The command inline, whose most simple   inline
synthax reads g=inline(expr,arg1,arg2,...,argn), declares a function g
which depends on the strings arg1,arg2,...,argn. The string expr contains
the expression of g. For instance, g=inline('sin(r)','r') declares the function $g(r) = \sin(r)$. The shorthand command g=inline(expr) implicitly assume that expr is a function of the default variable x. Once an inline function
has been declared, it can be evaluated at any set of variables through the
command feval. For instance, to evaluate g at the points z=[0 1] we can   feval
write

```
>> feval('g',z);
```

We note that, contrarily to the case of the command `eval`, with `feval` the name of the variable (`z`) needs not coincide with the symbolic name (`r`) assigned by the inline command.

After this quick introduction, our suggestion is to explore MATLAB using the command *help*, and get acquainted with the implementation of various algorithms by the programs described throughout this book.

See the Exercises 1.8-1.11.

1.7   What we haven't told you

A systematic discussion on floating-point numbers can be found in [Üeb97] or in [QSS00].

For matters concerning the issue of complexity, we refer, *e.g.*, to [Pan92].

For a more systematic introduction to MATLAB the interested reader can refer to the MATLAB manual [HH00] as well as to specific books such as [HLR01] or [EKH02].

1.8   Exercises

Exercise 1.1 How many numbers belong to the set $\mathbb{F}(2, 2, -2, 2)$? What is the value of ϵ_M for such set?

Exercise 1.2 Show that the set $\mathbb{F}(\beta, t, L, U)$ contains precisely $2(\beta-1)\beta^{t-1}(U - L + 1)$ elements.

Exercise 1.3 Write the MATLAB instructions to build an upper (respectively, lower) triangular matrix of dimension 10 having 2 on the main diagonal and -3 on the upper (respectively, lower) diagonal.

Exercise 1.4 Write the MATLAB instructions which allow the interchange of the third and seventh row of the matrices built up in Exercise 1.3, and then the instructions allowing the interchange between the fourth and eighth column.

Exercise 1.5 Verify whether the following vectors in $\mathbb{R}^4$ are linearly independent:

$$\mathbf{v}_1 = [0\ 1\ 0\ 1], \quad \mathbf{v}_2 = [1\ 2\ 3\ 4], \quad \mathbf{v}_3 = [1\ 0\ 1\ 0], \quad \mathbf{v}_4 = [0\ 0\ 1\ 1].$$

Exercise 1.6 Write the following functions and compute their first and second derivatives, as well as their primitives, using the symbolic toolbox of MATLAB:

$$g(x) = \sqrt{x^2 + 1}, \quad f(x) = \sin(x^3) + \cosh(x).$$

Exercise 1.7 For any given vector v of dimension n, using the command c=poly(v) one can construct the $n + 1$ coefficients of the polynomial $p(x) = $ poly $\sum_{k=1}^{n+1} c(k)x^{n+1-k}$ which is equal to $\Pi_{k=1}^{n}(x - v(k))$. In exact arithmetics, one should find that v = roots(poly(c)). However, this cannot occur due to roundoff errors, as one can check by using the command roots(poly([1:n])), where n ranges from 2 to 25.

Exercise 1.8 Write a program to compute the following sequence:

$$I_0 = \frac{1}{e}(e - 1),$$
$$I_{n+1} = 1 - (n + 1)I_n, \quad \text{for } n = 0, 1, \ldots,$$

Compare the numerical result with the exact limit $I_n \to 0$ for $n \to \infty$.

Exercise 1.9 Explain the behavior of the sequence (1.4) when computed in MATLAB.

Exercise 1.10 Consider the following algorithm to compute π. Generate n couples $\{(x_k, y_k)\}$ of random numbers in the interval $[0, 1]$, then compute the number m of them lying inside the first quarter of the unit circle. Obviously, π turns out to be the limit of the sequence $\pi_n = 4m/n$. Write a MATLAB program to compute this sequence and check the error for increasing values of n.

Exercise 1.11 Write a program for the computation of the binomial coefficient $\binom{n}{k} = n!/(k!(n - k)!)$, where n and k are two natural numbers with $k \leq n$.

2. Nonlinear equations

Computing the *zeros* of a function f (equivalently, the *roots* of the equation $f(x) = 0$) is a problem that we encounter quite often in scientific computing. In general, this task cannot be accomplished in a finite number of operations. For instance, we have already seen in Section 1.4.1 that when f is a generic polynomial of degree greater than 4, there do not exist closed formulae for the zeros. The situation is even more difficult when f is not a polynomial.

Iterative methods are therefore adopted. Starting from one or several initial data, the methods build up a sequence of values $x^{(k)}$ that hopefully will converge to a zero α of the function f at hand.

Problem 2.1 (Investment fund) At the beginning of every year a bank customer deposits v euros in an investment fund and withdraws, at the end of the n-th year, a capital of M euros. We want to compute the average yearly rate of interest I of this investment. Since M is related to I by the relation

$$M = v \sum_{k=1}^{n} (1+I)^k = v \frac{1+I}{I} [(1+I)^n - 1],$$

we deduce that I is the root of the algebraic equation:

$$f(I) = 0 \quad \text{where } f(I) = M - v\frac{1+I}{I}[(1+I)^n - 1].$$

This problem will be solved in Example 2.1. •

Problem 2.2 (State equation of a gas) We want to determine the volume V occupied by a gas at temperature T and pressure p. The state equation (i.e. the equation that relates p, V and T) is

$$[p + a(N/V)^2] (V - Nb) = kNT, \tag{2.1}$$

where a and b are two coefficients that depend on the specific gas, N is the number of molecules which are contained in the volume V and k is the Boltzmann constant. We need therefore to solve a nonlinear equation whose root is V (see Exercise 2.2). •

Problem 2.3 (Statics) Let us consider the mechanical system represented by the four rigid rods $\mathbf{a}_i$ of Figure 2.1. For any admissible value of the angle β, let us determine the value of the corresponding angle α between the rods $\mathbf{a}_1$ and $\mathbf{a}_2$. Starting from the vector identity

$$\mathbf{a}_1 - \mathbf{a}_2 - \mathbf{a}_3 - \mathbf{a}_4 = 0$$

and noting that the rod $\mathbf{a}_1$ is always aligned with the x-axis, we can deduce the following relationship between β and α:

$$\frac{a_1}{a_2}\cos(\beta) - \frac{a_1}{a_4}\cos(\alpha) - \cos(\beta - \alpha) = -\frac{a_1^2 + a_2^2 - a_3^2 + a_4^2}{2a_2a_4}, \tag{2.2}$$

where a_i is the known length of the i-th rod. This is called the Freudenstein equation, and we can rewrite it as follows: $f(\alpha) = 0$, where $f(x) = (a_1/a_2)\cos(\beta) - (a_1/a_4)\cos(x) - \cos(\beta - x) + (a_1^2 + a_2^2 - a_3^2 + a_4^2)/(2a_2a_4)$. A solution in explicit form is available only for special values of β. We would also like to mention that a solution does not exist for all values of β, and may not even be unique. To solve the equation for any given β lying between 0 and π we should invoke numerical methods (see Exercise 2.9). ●

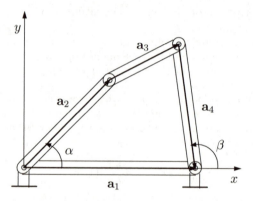

Fig. 2.1. System of four rods of Problem 2.3

2.1 The bisection method

Let f be a continuous function in $[a, b]$ which satisfies $f(a)f(b) < 0$. Then necessarily f has at least one zero in (a, b). Assume for simplicity that it is unique, and let us call it α.

(In the case of several zeros, by the help of the command `fplot` we can locate an interval which contains only one of them.)

The strategy of the bisection method is to halve the given interval and select that subinterval where f features a sign change (this subinterval

will henceforth contain α). This procedure is iterated until the last selected interval is so small that we have reached the desired accuracy. The
sequence $\{x^{(k)}\}$ of the midpoints of these subintervals $I^{(k)}$ will inevitably
tend to α since the length of the subintervals tends to zero as k tends to
infinity.

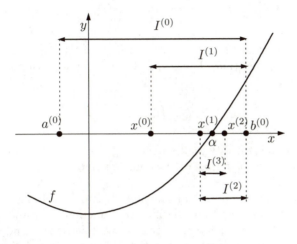

Fig. 2.2. A few iterations of the bisection method

Precisely, the method is started by setting

$$a^{(0)} = a, \quad b^{(0)} = b, \quad I^{(0)} = (a^{(0)}, b^{(0)}), \quad x^{(0)} = (a^{(0)} + b^{(0)})/2.$$

At the step $k \geq 1$ we select the subinterval $I^{(k)} = (a^{(k)}, b^{(k)})$ of the
interval $I^{(k-1)} = (a^{(k-1)}, b^{(k-1)})$ as follows:
given $x^{(k-1)} = (a^{(k-1)} + b^{(k-1)})/2$, if $f(x^{(k-1)}) = 0$ then $\alpha = x^{(k-1)}$
and the method terminates; otherwise,

if $f(a^{(k-1)})f(x^{(k-1)}) < 0$ we set $a^{(k)} = a^{(k-1)}$, $b^{(k)} = x^{(k-1)}$;

if $f(x^{(k-1)})f(b^{(k-1)}) < 0$ we set $a^{(k)} = x^{(k-1)}$, $b^{(k)} = b^{(k-1)}$.

Then we define $x^{(k)} = (a^{(k)} + b^{(k)})/2$ and k is increased by 1.

For instance, in the case represented in Figure 2.2, which corresponds
to the choice $f(x) = x^2 - 1$, by taking $a^{(0)} = -0.25$ and $b^{(0)} = 1.25$, we
would obtain

$$\begin{aligned}
I^{(0)} &= (-0.25, 1.25), & x^{(0)} &= 0.5, \\
I^{(1)} &= (0.5, 1.25), & x^{(1)} &= 0.875, \\
I^{(2)} &= (0.875, 1.25), & x^{(2)} &= 1.0625, \\
I^{(3)} &= (0.875, 1.0625), & x^{(3)} &= 0.96875.
\end{aligned}$$

Notice that every subinterval $I^{(k)}$ contains the zero α. Moreover, the sequence $\{x^{(k)}\}$ necessarily converges to α since at each step the length $|I^{(k)}| = b^{(k)} - a^{(k)}$ of $I^{(k)}$ halves. Since $|I^{(k)}| = (1/2)^k |I^{(0)}|$, the error at the $k - th$ step satisfies

$$|e^{(k)}| = |x^{(k)} - \alpha| < \frac{1}{2}|I^{(k)}| = \left(\frac{1}{2}\right)^{k+1} (b - a).$$

In order to guarantee that $|e^{(k)}| < \varepsilon$, for a given tolerance ε it suffices to carry out k_{min} iterations, being k_{min} the least integer that satisfies the inequality

$$k_{min} > \log_2 \left(\frac{b - a}{\varepsilon}\right) - 1. \tag{2.3}$$

Obviously, this inequality makes sense in general, and is not confined to the specific choice of f that we have made before.

The bisection method is implemented in Program 1: fun is a string (or an inline function) that specifies the function f, a and b are the extremes of the search interval, tol is the tolerance ε and nmax the maximum number of iterations allotted. When fun is an inline function (or a function defined in a m-file) besides the first argument relative to the independent function it can accept other auxiliary parameters needed for the definition of fun.

Output parameters are zero, which contains the approximate value of α, the residual res which is the value of f in zero and niter which is the total number of iterations that are carried out. The command find find(fx==0) finds those indices of the vector fx corresponding to null sign components, while the command sign(fx) determines the sign of fx.

Program 1 - bisection: bisection method

```
function [zero,res,niter]=bisection(fun,a,b,tol,nmax,varargin)
%BISECTION Find function zeros.
%   ZERO=BISECTION(FUN,A,B,TOL,NMAX) tries to find a zero ZERO of the
%   continuous function FUN in the interval [A,B] using the bisection method.
%   FUN accepts real scalar input x and returns a real scalar value. If the search
%   fails an errore message is displayed. FUN can also be an inline object.
%   ZERO=BISECTION(FUN,A,B,TOL,NMAX,P1,P2,...) passes parameters P1,
%   P2,... to the function FUN(X,P1,P2,...).
%   [ZERO,RES,NITER]= BISECTION(FUN,...) returns the value of the
%   residual in ZERO and the iteration number at which ZERO was computed.
x = [a, (a+b)*0.5, b];
fx = feval(fun,x,varargin{:});
```

```
if fx(1)*fx(3)>0
    error(' The sign of the function at the extrema of the interval must be different');
elseif fx(1) == 0
    zero = a; res = 0; niter = 0;
    return
elseif fx(3) == 0
    zero = b; res = 0; niter = 0;
    return
end
niter = 0;
l = (b - a)*0.5;
while l >= tol & niter <= nmax
    niter = niter + 1;
    if sign(fx(1))*sign(fx(2)) <  0
        x(3) = x(2);   x(2) = x(1)+(x(3)-x(1))*0.5;
        fx = feval(fun,x,varargin{:}); l = (x(3)-x(1))*0.5;
    elseif sign(fx(2))*sign(fx(3)) < 0
        x(1) = x(2);   x(2) = x(1)+(x(3)-x(1))*0.5;
        fx = feval(fun,x,varargin{:}); l = (x(3)-x(1))*0.5;
    else
        x(2) = x(find(fx==0)); l = 0;
    end
end
if niter > nmax
    fprintf(['bisection stopped without converging to the desired tolerance',...
    'because the maximum number of iterations was reached\n']);
end
zero = x(2); x = x(2); res = feval(fun,x);
```

Example 2.1 Let us apply the bisection method to solve Problem 2.1, assuming that v is equal to 1000 euros and that after 5 years M is equal to 6000 euros. The graph of the function f can be obtained by the following instructions

```
>> f=inline('M-v*(1+I).*((1+I).^5 - 1)./I','I','M','v');
>> fplot(f,[0.01,0.3],[],[],[],6000,1000)
```

We see that f has a unique zero in the interval $(0.01, 0.1)$, which is about equal to 0.06. If we execute the Program 1 with tol= 10^{-12}, a= 0.01 and b= 0.1, after 36 iterations the method converges to the value 0.061402, in perfect agreement with the estimate (2.3) according to which $k_{min} = 36$. We thus conclude that the interest rate I is approximately equal to 6.14%.

In spite of its simplicity, the bisection method does not guarantee a monotone reduction of the error, but simply that the search interval is halved from one iteration to the next. Consequently, if the only stopping criterion that is adopted is the control of the length of $I^{(k)}$, one might discard approximations of α which are quite accurate.

As a matter of fact, this method does not take into proper account the actual behavior of f. A striking fact is that it does not converge in a single iteration even if f is a linear function (unless the zero α is the midpoint of the initial search interval).

See Exercises 2.1-2.5.

2.2 The Newton method

A more efficient method for the computation of the zeros of a function f can be constructed by exploiting the differentiability of f. In that case,

$$y(x) = f(x^{(k)}) + f'(x^{(k)})(x - x^{(k)})$$

provides the equation of the tangent to the curve $(x, f(x))$ at the point $x^{(k)}$.

If we pretend that $x^{(k+1)}$ be such that $y(x^{(k+1)}) = 0$, we obtain:

$$x^{(k+1)} = x^{(k)} - \frac{f(x^{(k)})}{f'(x^{(k)})}, \text{ provided } f'(x^{(k)}) \neq 0 \qquad (2.4)$$

This formula allows us to compute a sequence of values $x^{(k)}$ starting from an initial guess $x^{(0)}$. This method is known as Newton's method and corresponds to computing the zero of f by locally replacing f by its tangent line (see Figure 2.3).

As a matter of fact, by developing f in Taylor series in a neighborhood of a generic point $x^{(k)}$ we find that

$$f(x^{(k+1)}) = f(x^{(k)}) + \delta^{(k)} f'(x^{(k)}) + \mathcal{O}((\delta^{(k)})^2), \qquad (2.5)$$

where $\delta^{(k)} = x^{(k+1)} - x^{(k)}$. Forcing $f(x^{(k+1)})$ to be zero and neglecting the term $\mathcal{O}((\delta^{(k)})^2)$, we can obtain $x^{(k+1)}$ as a function of $x^{(k)}$ as stated in (2.4). In this respect (2.4) can be regarded as an approximation of (2.5).

Obviously, (2.4) converges in a single step when f is linear, that is $f(x) = a_1 x + a_0$; in such a case

$$x^{(1)} = x^{(0)} - \frac{a_1 x^{(0)} + a_0}{a_1} = -\frac{a_0}{a_1}.$$

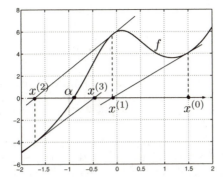

Fig. 2.3. The first iterations generated by the Newton method with initial guess $x^{(0)}$ for the function $f(x) = x + e^x + 10/(1 + x^2) - 5$

Example 2.2 Let us solve Problem 2.1 by Newton's method, taking as initial data $x^{(0)} = 0.3$. After 6 iterations the difference between two subsequent iterates is less than or equal to 10^{-12}.

The Newton method in general does not converge for all possible choices of $x^{(0)}$, but only for those values of $x^{(0)}$ which are *sufficiently close* to α. At a first glance, this request looks meaningless: indeed, in order to compute α (which is unknown), one should start from a value sufficiently close to α!

In practice, a possible initial value $x^{(0)}$ can be obtained by resorting to a few iterations of the bisection method or, alternatively, through an investigation of the graph of f. If $x^{(0)}$ is properly chosen and α is a simple zero (see Section 1.4.1) then the Newton method converges. Furthermore, in the special case in which f is continuously differentiable up to its second derivative one has the following convergence result (see Exercise 2.8),

$$\lim_{k \to \infty} \frac{x^{(k+1)} - \alpha}{(x^{(k)} - \alpha)^2} = \frac{f''(\alpha)}{2f'(\alpha)} \tag{2.6}$$

Consequently, Newton's method is said to converge *quadratically*, or with order 2, since for sufficiently large values of k the error at the $(k+1)$-th step behaves like the square of the error at the k-th step multiplied by a constant which is independent of k.

In the case of zeros with multiplicity m larger than 1, the order of convergence of Newton's method downgrades to 1 (see Exercise 2.15). In that case one could recover the order 2 by modifying the original method as follows:

$$x^{(k+1)} = x^{(k)} - m\frac{f(x^{(k)})}{f'(x^{(k)})}, \text{ provided } f'(x^{(k)}) \neq 0. \qquad (2.7)$$

Obviously, this *modified Newton method* requires the *a-priori* knowledge of m. If this is not the case, one could develop an *adaptive Newton method*, still of order 2, as described in [QSS00], Section 6.6.2.

Example 2.3 The function $f(x) = (x-1)\log(x)$ has a single zero $\alpha = 1$ of multiplicity $m = 2$. Let us compute it by both Newton's method (2.4) and by its modified version (2.7). In Figure 2.4 we report the error obtained using the two methods versus the iteration number. Note that for the classical Newton's method the convergence is only linear.

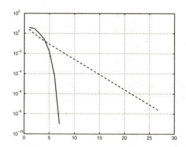

Fig. 2.4. Error versus iteration number for the function of Example 2.3. Dashed line corresponds to Newton's method (2.4), solid line to the modified Newton's method (2.7) (with $m = 2$)

In theory, a convergent Newton's method returns the zero α only after an infinite number of iterations. In practice, one requires an approximation of α up to a prescribed tolerance ε. Thus the iterations can be terminated at the least k_{min} for which the following inequality holds:

$$\left|e^{(k_{min})}\right| = \left|\alpha - x^{(k_{min})}\right| < \varepsilon.$$

Unfortunately, since the error is unknown, one needs to employ in its place a suitable *error estimator*, that is, a quantity that can be easily computed and through which we can estimate the real error. At the end of Section 2.3, we will see that a suitable error estimator for Newton's method is provided by the difference between two successive iterates. This means that one terminates the iterations at the k_{min}-th step as soon as

$$\left|x^{(k_{min})} - x^{(k_{min}-1)}\right| < \varepsilon. \qquad (2.8)$$

Remark 2.1 (The general case) The stopping criterion (2.8) may not be suitable for methods different from Newton's. Alternatively, one could use an error estimator based on the *residual* at the step k defined as $r^{(k)} = f(x^{(k)})$ (note that the residual is null when $x^{(k)}$ is a zero of the function f).

We could stop the iteration at the first k_{min} for which

$$|f(x^{(k_{min})})| < \varepsilon.$$

Using the residual as an error estimator is satisfactory only when the behavior of f is almost linear in a neighborhood I_α of the zero α (see Figure 2.5). Otherwise, it will produce an over estimation of the error if $|f'(x)| \gg 1$ for $x \in I_\alpha$ and an under estimation if $|f'(x)| \ll 1$ (see also Exercise 2.6).

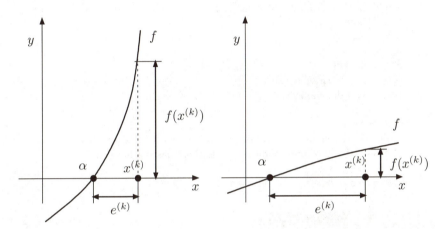

Fig. 2.5. Two situations in which the residual is a poor error estimator: $|f'(x)| \gg 1$ (left), $|f'(x)| \ll 1$ (right), with x belonging to a neighborhood of α

In Program 2 we implement Newton's method (2.4). Its modified form can be obtained simply by replacing f' with f'/m. The input parameters f and df are the strings which define the function f and its first derivative, while x0 is the initial guess. The method will be terminated when the absolute value of the difference between two subsequent iterates is less than the prescribed tolerance tol, or when the maximum number of iterations nmax has been reached.

Program 2 - newton: Newton's method

```
function [zero,res,niter]=newton(f,df,x0,tol,nmax,varargin)
%NEWTON Find function zeros.
%  ZERO=NEWTON(FUN,DFUN,X0,TOL,NMAX) tries to find the zero ZERO of
%  the continuous and differentiable function FUN nearest to X0 using the Newton
```

```
%   method. FUN and its derivative DFUN accept real scalar input x and returns
%   a real scalar value. If the search fails an errore message is displayed.
%   FUN and DFUN can also be inline objects.
%   ZERO=NEWTON(FUN,DFUN,X0,TOL,NMAX,P1,P2,...) passes parameters
%   P1,P2,... to functions: FUN(X,P1,P2,...) and DFUN(X,P1,P2,...).
%   [ZERO,RES,NITER]= NEWTON(FUN,...) returns the value of the
%   residual in ZERO and the iteration number at which ZERO was computed.
x = x0;
fx = feval(f,x,varargin{:});
dfx = feval(df,x,varargin{:});
niter = 0;
diff = tol+1;
while diff >= tol & niter <= nmax
    niter = niter + 1;
    diff = - fx/dfx;
    x = x + diff;
    diff = abs(diff);
    fx = feval(f,x,varargin{:});
    dfx = feval(df,x,varargin{:});
end
if niter > nmax
    fprintf(['newton stopped without converging to the desired tolerance',...
    'because the maximum number of iterations was reached\n']);
end
zero = x; res = fx;
```

Let us summarize

1. Methods for the computation of the zeros of a function f are usually of iterative type;

2. the bisection method computes a zero of a function f by generating a sequence of intervals whose length is halved at each iteration. This method is convergent provided that f is continuous in the initial interval and has opposite signs at the end-points of this interval;

3. Newton's method computes a zero α of f by taking into account the values of f and of its derivative. A necessary condition for convergence is that the initial datum belongs to a suitable (sufficiently small) neighborhood of α;

4. Newton's method is quadratically convergent only when α is a simple zero of f, otherwise convergence is linear.

See Exercises 2.6-2.14.

2.3   Fixed point iterations

Playing with a pocket calculator, one may verify that by applying repeatedly the cosine key to the real value 1, one gets the following sequence of real numbers:

$$x^{(1)} = \cos(1) = 0.54030230586814,$$
$$x^{(2)} = \cos(x^{(1)}) = 0.85755321584639,$$
$$x^{(10)} = \cos(x^{(9)}) = 0.74423735490056,$$
$$x^{(20)} = \cos(x^{(19)}) = 0.73918439977149,$$

which should tend to the value $\alpha = 0.73908513\ldots$. Since, by construction, $x^{(k+1)} = \cos(x^{(k)})$ for $k = 0, 1, \ldots$ (with $x^{(0)} = 1$), the limit α satisfies the equation $\cos(\alpha) = \alpha$. For this reason α is called a fixed point of the cosine function. We may wonder how such iterations could be exploited in order to compute the zeros of a given function. In the previous example, α is not only a fixed point for the cosine function, but also a zero of the function $f(x) = x - \cos(x)$, hence the method previously proposed can be regarded as a method to compute the zeros of f. On the other hand, not every function has fixed points. For instance, by repeating the previous experiment using now the exponential function and still $x^{(0)} = 1$ one encounters a situation of overflow after 4 steps only (see Figure 2.6).

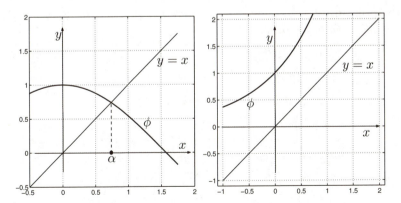

Fig. 2.6. The function $\phi(x) = \cos x$ admits one and only one fixed point (left), whereas the function $\phi(x) = e^x$ does not have any (right)

Let us clarify the above intuitive idea by considering the following problem. Given a function $\phi : [a, b] \to \mathbb{R}$, find $\alpha \in [a, b]$ such that

$$\alpha = \phi(\alpha).$$

If such an α exists it will be called a fixed point of ϕ and it could be computed by the following algorithm:

$$x^{(k+1)} = \phi(x^{(k)}), \quad k \geq 0 \qquad (2.9)$$

where $x^{(0)}$ is an initial guess. This algorithm is called *fixed point iterations* and ϕ is said to be the *iteration function*. The introductory example is therefore an instance of fixed point iterations with $\phi(x) = \cos(x)$.

A geometrical interpretation of (2.9) is provided in Figure 2.7 (left). One can guess that if ϕ is a continuous function and the limit of the sequence $\{x^{(k)}\}$ exists, then such limit is a fixed point of ϕ. We will make this result more precise in Proposition 2.1 and 2.2.

Example 2.4 The Newton method (2.4) can be regarded as an algorithm of fixed point iterations whose iteration function is

$$\phi(x) = x - \frac{f(x)}{f'(x)}. \qquad (2.10)$$

From now on this function will be denoted by ϕ_N (where N stands for Newton). This is not the case for the bisection method since the generic iterate $x^{(k+1)}$ depends not only on $x^{(k)}$ but also on $x^{(k-1)}$.

As shown in Figure 2.7 (right), fixed point iterations may not converge. Indeed, the following result holds.

Proposition 2.1 *Assume that the iteration function in (2.9) satisfies the following properties:*

1. *$\phi(x) \in [a, b]$ for all $x \in [a, b]$;*

2. *ϕ is differentiable in $[a, b]$;*

3. *$\exists K < 1$ such that $|\phi'(x)| \leq K$ for all $x \in [a, b]$.*

Then ϕ has a unique fixed point $\alpha \in [a, b]$ and the sequence defined in (2.9) converges to α, whatever choice is made for the initial datum $x^{(0)}$ in $[a, b]$. Moreover

$$\lim_{k \to \infty} \frac{x^{(k+1)} - \alpha}{x^{(k)} - \alpha} = \phi'(\alpha) \qquad (2.11)$$

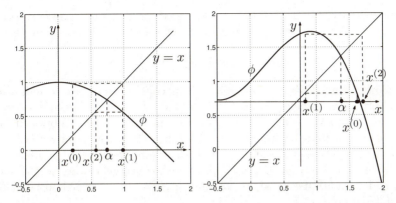

Fig. 2.7. Representation of a few fixed point iterations for two different iteration functions. To the left, the iterations converge to the fixed point α, whereas the iterations on the right produce a divergence sequence

From (2.11) one deduces that the fixed point iterations converge at least linearly, that is, for k sufficiently large the error at the step $k+1$ behaves like that at the k step multiplied by a constant $\phi'(\alpha)$ which is independent of k and whose absolute value is strictly less than 1.

Example 2.5 The function $\phi(x) = \cos(x)$ satisfies all the assumptions of Proposition 2.1. Indeed, $|\phi'(\alpha)| = |\sin(\alpha)| \simeq 0.67 < 1$, and thus by continuity there exists a neighborhood I_α of α such that $|\phi'(x)| < 1$ for all $x \in I_\alpha$. The function $\phi(x) = x^2 - 1$, although possessing two fixed points $\alpha_{\pm} = (1 \pm \sqrt{5})/2$, does not satisfy the assumption for either since $|\phi'(\alpha_{\pm})| = |1 \pm \sqrt{5}| > 1$. The corresponding fixed point iterations will not converge.

The Newton method is not the only iterative procedure featuring quadratic convergence. Indeed, the following general property holds.

Proposition 2.2 *Assume that all hypotheses of Proposition 2.1 are satisfied. In addition assume that ϕ is differentiable twice and that*

$$\phi'(\alpha) = 0, \quad \phi''(\alpha) \neq 0.$$

Then the fixed point iterations (2.9) converge with order 2 and

$$\lim_{k \to \infty} \frac{x^{(k+1)} - \alpha}{(x^{(k)} - \alpha)^2} = \frac{1}{2}\phi''(\alpha). \tag{2.12}$$

Example 2.4 shows that the fixed point iterations (2.9) could also be used to compute the zeros of the function f. Clearly for any given f the function ϕ defined in (2.10) is not the only possible iteration function.

For instance, for the solution of the equation $\log(x) = \gamma$, after setting $f(x) = \log(x) - \gamma$, the choice (2.10) could lead to the iteration function

$$\phi_N(x) = x(1 - \log(x) + \gamma).$$

Another fixed point iteration algorithm could be obtained by adding x to both sides of the equation $f(x) = 0$. The associated iteration function is now $\phi_1(x) = x + \log(x) - \gamma$. A further method could be obtained by choosing the iteration function $\phi_2(x) = x \log(x)/\gamma$. Not all these methods are convergent. For instance, if $\gamma = -2$, the methods corresponding to the iteration functions ϕ_N and ϕ_2 are both convergent, whereas the one with iteration function ϕ_1 is not since $|\phi_1'(x)| > 1$ in a neighborhood of the fixed point α.

2.3.1   How to terminate fixed point iterations

In general, fixed point iterations are terminated when the absolute value of the difference between two consecutive iterates is less than a prescribed tolerance ε.

Since $\alpha = \phi(\alpha)$ and $x^{(k+1)} = \phi(x^{(k)})$, using the mean value theorem (see Section 1.4.3) we find

$$\alpha - x^{(k+1)} = \phi(\alpha) - \phi(x^{(k)}) = \phi'(\xi^{(k)})\,(\alpha - x^{(k)}) \text{ with } \xi^{(k)} \in I_{\alpha,x^{(k)}},$$

$I_{\alpha,x^{(k)}}$ being the interval with end-points α and $x^{(k)}$. Using the identity

$$\alpha - x^{(k)} = (\alpha - x^{(k+1)}) + (x^{(k+1)} - x^{(k)}),$$

it follows that

$$\alpha - x^{(k)} = \frac{1}{1 - \phi'(\xi^{(k)})}(x^{(k+1)} - x^{(k)}). \tag{2.13}$$

Consequently, if $\phi'(x) \simeq 0$ in a neighborhood of α, the difference between two consecutive iterates provides a satisfactory error estimator. This is the case for methods of order 2, including Newton's method. This estimate becomes as more unsatisfactory as ϕ' approaches 1.

Example 2.6 Let us compute with Newton's method the zero $\alpha = 1$ of the function $f(x) = (x-1)^{m-1}\log(x)$ for $m = 11$ and $m = 21$, whose multiplicity is equal to m. In this case Newton's method converges with order 1; moreover, it is possible to prove (see Exercise 2.15) that $\phi_N'(\alpha) = 1 - 1/m$, ϕ_N being the iteration function of the method, regarded as a fixed point iteration algorithm. As m increases, the accuracy of the error estimate furnished by the difference between two consecutive iterates decreases. This is confirmed by the numerical results in Figure 2.8 where we compare the behavior of the true error with that of our estimator for both $m = 11$ and $m = 21$. The difference between these two quantities increases as m increases.

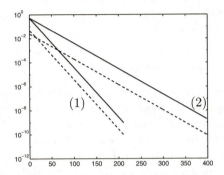

Fig. 2.8. Absolute values of the errors (solid line) and absolute values of the difference between two consecutive iterates (dashed line), plotted versus the number of iterations for the case of Example 2.6. Graphs (1) refer to $m = 11$, graphs (2) to $m = 21$

Let us summarize

1. A number α satisfying $\phi(\alpha) = \alpha$ is called a fixed point of ϕ. For its computation we can use the so-called fixed point iterations: $x^{(k+1)} = \phi(x^{(k)})$;

2. fixed point iterations converge under suitable assumptions on ϕ and its first derivative. Typically, convergence is linear, however, in the special case when $\phi'(\alpha) = 0$, the fixed point iterations converge quadratically;

3. fixed point iterations can be used also to compute the zeros of a function.

See Exercises 2.15-2.18.

2.4 What we haven't told you

The most sophisticated methods for the computation of the zeros of a function combine different algorithms. In particular, the MATLAB func-

fzero tion **fzero** (see Section 1.4.1) adopts the so called Dekker-Brent method (see [QSS00], Section 6.2.3). In its basic form **fzero(fun,x0)** computes the zero of the function **fun**, where **fun** can be either a string which is function of **x**, or the name of an inline function or the name of a m-file.

For instance, we could solve the problem of Example 2.1 also by **fzero**, using the initial value **x0=0.3** (as done by Newton's method) through the following instructions:

```
>> f=inline('6000 - 1000*(1+I)/I*((1+I)^5 - 1)','I'); x0=0.3;
>> [alpha,res,flag,iter]=fzero(f,x0);
```

we obtain **alpha=0.06140241153653** with residual **res=9.0949e-13** in **iter=29** iterations. When **flag** is negative it means that **fzero** cannot find the zero. The Newton method converges in 6 iterations to the value 0.06140241153652 with a residual equal to 2.3646e-11.

In this chapter, we have frequently mentioned the problem of how to compute the zeros of a polynomial. The solution of this problem requires *ad hoc* algorithms which are typically based on a *deflation* technique, a procedure which eliminates automatically the zeros already computed. Among others let us mention the Sturm method, the Müller method, the Newton-Hörner method (see [Atk89] or [QSS00]) and the Bairstow method ([RR85]). A different approach consists of characterizing the zeros of a function as the eigenvalues of a special matrix (called the *companion matrix*) and then using appropriate techniques for their computation. This approach is adopted by the MATLAB function **roots** which has been introduced in Section 1.4.2.

Newton's method and fixed point iterations can be easily extended to compute the roots of nonlinear systems of the following form (see [QSS00], Chapter 7):

$$
\begin{cases}
f_1(x_1, x_2, \ldots, x_n) = 0, \\
f_2(x_1, x_2, \ldots, x_n) = 0, \\
\vdots \\
f_n(x_1, x_2, \ldots, x_n) = 0,
\end{cases}
$$

where $f_1, \ldots, f_n$ are nonlinear functions. Other methods exist as well, such as the Broyden and quasi-Newton methods, and can be regarded as generalizations of Newton's method (see [DS83]).

The MATLAB instruction

$$\texttt{zero} = \texttt{fsolve}('\texttt{fun}', \texttt{x0})$$

allows the computation of one zero of a nonlinear system defined through the user function **fun** starting from the vector **x0** as initial guess. The function **fun** returns the n values $f_1(x), \ldots, f_n(x)$ for any value of the input vector x.

For instance, let us consider the following system:

$$\begin{cases} x^2 + y^2 = 1, \\ \sin(\pi x/2) + y^3 = 0, \end{cases} \qquad (2.14)$$

whose solutions are $(0.4761, -0.8794)$ and $(-0.4761, 0.8794)$.

The corresponding MATLAB user function, which we call **systemnl**, is defined as follows:

```
function fx=systemnl(x)
fx(1) = x(1)^2+x(2)^2-1;
fx(2) = sin(pi*0.5*x(1))+x(2)^3;
```

The MATLAB instructions to solve this system are therefore:

```
>> x0 = [1 1];
>> alpha=fsolve('systemnl',x0)
alpha =
   0.4761   -0.8794
```

Using this procedure we have found only one of the two roots. The other can be computed starting from the initial datum -x0.

2.5 Exercises

Exercise 2.1 Given the function $f(x) = \cosh x + \cos x - \gamma$, for $\gamma = 1, 2, 3$ find an interval that contains the zero of f. Then compute the zero by the bisection method with a tolerance of 10^{-10}.

Exercise 2.2 For carbon dioxide (CO_2) the coefficients a and b in (2.1) take the following values: $a = 0.401 Pa\ m^6$, $b = 42.7 \cdot 10^{-6} m^3$ (Pa stands for Pascal). Find the volume occupied by 1000 molecules of CO_2 at a temperature $T = 300K$ and a pressure $p = 3.5 \cdot 10^7 Pa$ by the bisection method, with a tolerance of 10^{-12} (the Boltzmann constant is $k = 1.3806503 \cdot 10^{-23} Joule\,K^{-1}$).

Exercise 2.3 An object is standing on a plane whose slope varies with constant velocity ω. After t seconds its position is

$$s(t, \omega) = \frac{g}{2\omega^2} [\sinh(\omega t) - \sin(\omega t)],$$

where $g = 9.8 m/s^2$ denotes the gravity acceleration. Assuming that this object has moved by 1 meter in 1 second, compute the corresponding value of ω with a tolerance of 10^{-5}.

Exercise 2.4 Prove inequality (2.3).

Exercise 2.5 Motivate why in Program 1 the instruction `x(2) = x(1)+(x(3)-x(1))*0.5` has been used instead of the more natural one `x(2)=(x(1)+x(3))*0.5` in order to compute the midpoint.

Exercise 2.6 Apply Newton's method to solve Exercise 2.1. Why is this method not accurate when $\gamma = 2$?

Exercise 2.7 Apply Newton's method to compute the square root of a positive number a. Proceed in a similar manner to compute the cube root of a.

Exercise 2.8 Assuming that Newton's method converges, show that (2.6) is true when α is a simple root of $f(x) = 0$ and f is twice continuously differentiable in a neighborhood of α.

Exercise 2.9 Apply Newton's method to solve Problem 2.3 for $\beta \in [0, 2\pi/3]$ with a tolerance of 10^{-5}. Assume that the lengths of rods are $a_1 = 10$ cm, $a_2 = 13$ cm, $a_3 = 8$ cm and $a_4 = 10$ cm. For each value of β consider two possible initial data, $x^{(0)} = -0.1$ and $x^{(0)} = 2\pi/3$.

Exercise 2.10 Notice that the function $f(x) = e^x - 2x^2$ has 3 zeros, $\alpha_1 < 0$, α_2 and α_3 positive. For which value of $x^{(0)}$ does Newton's method converge to α_1?

Exercise 2.11 Use Newton's method to compute the zero of $f(x) = x^3 - 3x^2 2^{-x} + 3x4^{-x} - 8^{-x}$ in $[0, 1]$ and explain why convergence is not quadratic.

Exercise 2.12 A projectile is ejected with velocity v_0 and angle α in a tunnel of height h and reaches its maximum range when α is such that $\sin(\alpha) = \sqrt{2gh/v_0^2}$, where $g = 9.8$ m/s^2 is the gravity acceleration. Compute α using Newton's method, assuming that $v_0 = 10$ m/s and $h = 1$ m.

Exercise 2.13 Solve Problem 2.1 by Newton's method with a tolerance of 10^{-12}, assuming $M = 6000$ euros, $v = 1000$ euros and $n = 5$. As an initial guess take the result obtained after 5 iterations of the bisection method applied on the interval $(0.01, 0.1)$.

Exercise 2.14 A corridor has the form indicated in Figure 2.9. The maximum length L of a rod that can pass from one extreme to the other sliding on the ground is given by

$$L = l_2/(\sin(\pi - \gamma - \alpha)) + l_1/\sin(\alpha),$$

where α is the solution of the nonlinear equation

$$l_2 \frac{\cos(\pi - \gamma - \alpha)}{\sin^2(\pi - \gamma - \alpha)} - l_1 \frac{\cos(\alpha)}{\sin^2(\alpha)} = 0. \qquad (2.15)$$

Compute α by Newton's method when $l_2 = 10$, $l_1 = 8$ and $\gamma = 3\pi/5$.

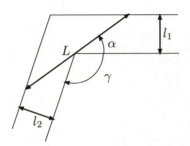

Fig. 2.9. The problem of a rod sliding in a corridor

Exercise 2.15 Let ϕ_N be the iteration function of Newton's method when regarded as a fixed point iteration. Show that $\phi_N'(\alpha) = 1 - 1/m$ where α is a zero of f with multiplicity m. Deduce that Newton's method converges quadratically if α is a simple root of $f(x) = 0$, and linearly otherwise.

Exercise 2.16 Deduce from the graph of $f(x) = x^3 + 4x^2 - 10$ that this function has a unique real zero α. To compute α use the following fixed point iterations: given $x^{(0)}$, define $x^{(k+1)}$ such that

$$x^{(k+1)} = \frac{2(x^{(k)})^3 + 4(x^{(k)})^2 + 10}{3(x^{(k)})^2 + 8x^{(k)}}, \qquad k \geq 0$$

and analyze its convergence to α.

Exercise 2.17 Analyze the convergence of the fixed point iterations

$$x^{(k+1)} = \frac{x^{(k)}[(x^{(k)})^2 + 3a]}{3(x^{(k)})^2 + a}, \qquad k \geq 0,$$

for the computation of the square root of a positive number a.

Exercise 2.18 Repeat the computations carried out in Exercise 2.11 using now the stopping criterion based on the residual (see Remark 2.1). Which is the more accurate result?

3. Approximation of functions and data

Approximating a function f consists of replacing it by another function $\tilde{f}$ of simpler form that may be used as its surrogate. This strategy is used frequently in numerical integration where, instead of computing $\int_a^b f(x)dx$, one carries out the exact computation of $\int_a^b \tilde{f}(x)dx$, $\tilde{f}$ being a function simple to integrate (e.g. a polynomial), as we will see in the next chapter. In other instances the function f may be available only partially through its values at some selected points. In these cases we aim at constructing a continuous function $\tilde{f}$ that could represent the empirical law which is behind the finite set of data. We provide a couple of examples which illustrate this kind of approach.

Problem 3.1 (Climatology) The air temperature near the ground depends on the concentration K of the carbon acid therein. In Table 3.1 (taken from Philosophical Magazine 41, 237 (1896)) we report for 3 different values of K the variation of the average temperature with respect to the actual average temperature (normalized at the reference value $K = 1$) at different latitudes on the Earth. In this case we can generate a function that, on the basis of the available data, provides an approximate value of the average temperature at any possible latitude and for other values of K (see Example 3.1). •

Problem 3.2 (Finance) In Figure 3.1 we report the price of a stock at the Zurich stock exchange over two years. The curve was obtained by joining with a straight line the prices reported at every day's closure. This simple representation indeed implicitly assumes that the prices change linearly in the course of the day (we anticipate that this approximation is called composite linear interpolation). We ask wheter from this graph one could predict the stock price for a short time interval beyond the time of the last quotation. We will see in Section 3.4 that this kind of prediction could be guessed by

Latitude	$K = 0.67$	$K = 1.5$	$K = 2.0$	$K = 3.0$
65	-3.1	3.52	6.05	9.3
55	-3.22	3.62	6.02	9.3
45	-3.3	3.65	5.92	9.17
35	-3.32	3.52	5.7	8.82
25	-3.17	3.47	5.3	8.1
15	-3.07	3.25	5.02	7.52
5	-3.02	3.15	4.95	7.3
-5	-3.02	3.15	4.97	7.35
-15	-3.12	3.2	5.07	7.62
-25	-3.2	3.27	5.35	8.22
-35	-3.35	3.52	5.62	8.8
-45	-3.37	3.7	5.95	9.25
-55	-3.25	3.7	6.1	9.5

Tab. 3.1. Variation of the average yearly temperature on the Earth for different values of the concentration K of carbon acid at different latitudes

resorting to a special technique known as *least squares* approximation of data (see Example 3.8). •

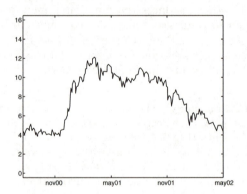

Fig. 3.1. Price variation of a stock over two years

It is known that a function f can be successfully replaced in a given interval by its Taylor polynomial, which was introduced in Section 1.4.3. This technique is computationally expensive since it requires the knowledge of f and its derivatives up to the order n (the polynomial degree) at a given point x_0. A further drawback is represented by the fact that the Taylor polynomial may fail to accurately represent f far enough from the point x_0. For instance, in Figure 3.2 we compare the behavior of $f(x) = 1/x$ with that of its Taylor polynomial of degree 10 built around the point $x_0 = 1$. This picture also shows the graphical interface of the

MATLAB function `taylortool` which allows the computation of Taylor's   `taylortool`
polynomial of arbitrary degree for any given function f. The agreement
between the function and its Taylor polynomial is very good in a small
neighborhood of $x_0 = 1$ while it becomes unsatisfactory when $x - x_0$
gets large. Fortunately, this is not the case of other functions such as the
exponential function which is approximated quite nicely for all $x \in \mathbb{R}$ by
its Taylor polynomial related to $x_0 = 0$, provided that the degree n is
sufficiently large.

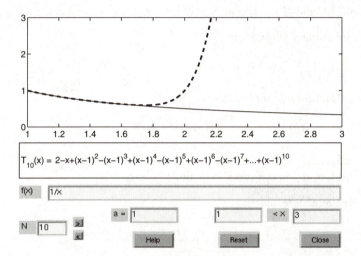

Fig. 3.2. Comparison between the function $f(x) = 1/x$ (solid line) and its
Taylor polynomial of degree 10 related to the point $x_0 = 1$ (dashed line). The
explicit form of the Taylor polynomial is also reported

In the course of this chapter we will introduce approximation methods
that are based on alternative approaches.

3.1 Interpolation

As seen in Problems 3.1 and 3.2, in several applications it may happen
that a function is known only through its values at some given points.
We are therefore facing a (general) case where $n + 1$ couples $\{x_i, f(x_i)\}$,
$i = 0, \ldots, n$, are given; the points x_i are all distinct and are called *nodes*.
For instance in the case of Table 3.1, n is equal to 12, the nodes x_i are
the values of the latitude reported in the first column, while the $f(x_i)$ are
the corresponding values (of the temperature) in the remaining columns.

In such a situation it seems natural to require the approximate function $\tilde{f}$ to satisfy the set of relations

$$\tilde{f}(x_i) = f(x_i), \quad i = 0, 1, \ldots, n \tag{3.1}$$

Such an $\tilde{f}$ is called *interpolant* of f and equations (3.1) are the interpolation conditions.

Several kinds of interpolants could be envisaged, such as:

- *polynomial interpolant*:

$$\tilde{f}(x) = a_0 + a_1 x + a_2 x^2 + \ldots + a_n x^n;$$

- *trigonometric interpolant*:

$$\tilde{f}(x) = a_{-M} e^{-iMx} + \ldots + a_0 + \ldots + a_M e^{iMx}$$

where M is an integer equal to $n/2$ if n is even, $(n-1)/2$ if n is odd, and i is the imaginary unit;

- *rational interpolant*:

$$\tilde{f}(x) = \frac{a_0 + a_1 x + \ldots + a_k x^k}{a_{k+1} + a_{k+2} x + \ldots a_{k+n-1} x^n}.$$

For simplicity we only consider those interpolants which depend linearly on the $n+1$ unknown coefficients a_i. Both polynomial and trigonometric interpolation fall into this category, whereas the rational interpolant does not.

3.1.1 Lagrangian polynomial interpolation

Let us focus on the polynomial interpolation. The following result holds:

Proposition 3.1 *For any set of couples* $\{x_i, f(x_i)\}$, $i = 0, \ldots, n$, *with distinct nodes* x_i, *there exists a unique polynomial of degree less than or equal to* n, *which we indicate by* $\Pi_n f$ *and call interpolating polynomial of the values* $f(x_i)$ *at the nodes* x_i, *such that*

$$\Pi_n f(x_i) = f(x_i), \quad i = 0, \ldots, n \tag{3.2}$$

In the case where the $f(x_i)$, $i = 0, \ldots, n$, represent the values attained by a continuous function f, $\Pi_n f$ is called interpolating polynomial of f (in short, interpolant of f).

To verify uniqueness we proceed by contradiction and suppose that there exist two polynomials of degree n, $\Pi_n f$ and $\Pi'_n f$, both satisfying the nodal relation (3.2). Their difference, $\Pi_n f - \Pi'_n f$, would be a polynomial of degree n which vanishes at $n + 1$ distinct points. Owing to a well known theorem of Algebra, such a polynomial should vanish identically, and then $\Pi'_n f$ must coincide with $\Pi_n f$.

In order to obtain an expression for $\Pi_n f$, we start from a very special case where $f(x_i)$ vanishes for all i apart from $i = k$ (for a fixed k) for which $f(x_k) = 1$. Then setting $\varphi_k(x) = \Pi_n f(x)$, we must have (see Figure 3.3)

$$\varphi_k \in \mathbb{P}_n, \quad \varphi_k(x_j) = \delta_{jk} = \begin{cases} 1 & \text{if } j = k, \\ 0 & \text{otherwise} \end{cases}$$

(δ_{jk} is the Kronecker symbol).

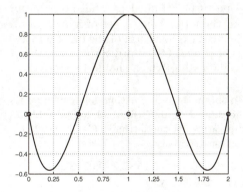

Fig. 3.3. The polynomial $\varphi_2 \in \mathbb{P}_4$ associated with a set of 5 equispaced nodes

The functions φ_k can be written as follows:

$$\varphi_k(x) = \prod_{\substack{j=0 \\ j \neq k}}^{n} \frac{x - x_j}{x_k - x_j}, \quad k = 0, \ldots, n. \tag{3.3}$$

We move now to the general case where $\{f(x_i), i = 0, \ldots, n\}$ is a set of arbitrary values. Using an obvious superposition principle we can obtain the following expression for $\Pi_n f$:

$$\Pi_n f(x) = \sum_{k=0}^{n} f(x_k)\varphi_k(x) \tag{3.4}$$

Indeed, this polynomial satisfies the interpolation conditions (3.2), since

$$\Pi_n f(x_i) = \sum_{k=0}^{n} f(x_k)\varphi_k(x_i) = \sum_{k=0}^{n} f(x_k)\delta_{ik} = f(x_i), \quad i = 0, \dots, n.$$

Due to their special role, the functions φ_k are called *Lagrange characteristic polynomials*, and (3.4) is the *Lagrange form* of the interpolant.

polyfit   In MATLAB we can store the n+1 couples $\{(x_i, f(x_i))\}$ in the vectors x and y, and then the instruction c=polyfit(x,y,n) will provide the coefficients of the interpolating polynomial. Precisely, c(1) will contain the coefficient of x^n, c(2) that of x^{n-1}, ... and c(n+1) the value of $\Pi_n f(0)$. (More on this command can be found in Section 3.4.) As already seen in Chapter 1, we can then use the instruction p=polyval(c,z) to compute the value p(j) attained by the interpolating polynomial at z(j), j=1,...,m, the latter being a set of m arbitrary points.

In the case when the explicit form of the function f is available, we can use the instruction y=eval(f) in order to obtain the vector y of values of f at some specific nodes (which should be stored in a vector x).

Example 3.1 To obtain the interpolating polynomial for the data of Problem 3.1 relating to the value $K = 0.67$ (first column of Table 3.1), using only the values of the temperature for the latitudes 65, 35, 5, -25, -55, we can use the following MATLAB instructions:

```
>> x = [-55 -25 5 35 65]; y = [-3.25 -3.2 -3.02 -3.32 -3.1];
>> format short e; c=polyfit(x,y,4)
c =
   8.2819e-08  -4.5267e-07  -3.4684e-04   3.7757e-04  -3.0132e+00
```

The graph of the interpolating polynomial can be obtained as follows:

```
>> z=linspace(x(1),x(end),100);
>> p=polyval(c,z); plot(z,p,x,y,'o');
```

In order to get a smooth curve we have evaluated our polynomial at 101 equispaced points in the interval $[-55, 65]$ (as a matter of fact, MATLAB plots are always constructed on piecewise linear interpolation between neighboring points). Note that the instruction x(end) picks up directly the last component of the vector x, without specifying the length of the vector. In Figure 3.4 the filled circles correspond to those values which have been used to construct the interpolating polynomial, whereas the empty circles correspond to values that have not been used. We can appreciate the qualitative agreement between the curve and the data distribution.

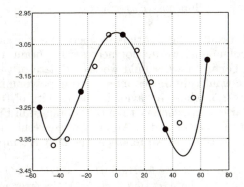

Fig. 3.4. The interpolating polynomial of degree 4 introduced in Example 3.1

Using the following result we can evaluate the error obtained by replacing f with its interpolating polynomial $\Pi_n f$:

Proposition 3.2 *Let I be a bounded interval, and consider $n + 1$ distinct interpolation nodes $\{x_i, i = 0, \ldots, n\}$ in I. Let f be continuously differentiable up to order $n + 1$ in I. Then $\forall x \in I \; \exists \xi \in I$ such that*

$$E_n f(x) = f(x) - \Pi_n f(x) = \frac{f^{(n+1)}(\xi)}{(n+1)!} \prod_{i=0}^{n} (x - x_i) \qquad (3.5)$$

Obviously, $E_n f(x_i) = 0$, $i = 0, \ldots, n$.

Result (3.5) can be better specified in the case of a uniform distribution of nodes, that is when $x_i = x_{i-1} + h$ for $i = 1, \ldots, n$, for a given $h > 0$ and a given x_0. As stated in Exercise 3.1, $\forall x \in (x_0, x_n)$ one can verify that

$$\left| \prod_{i=0}^{n} (x - x_i) \right| \le n! \frac{h^{n+1}}{4}, \qquad (3.6)$$

and therefore

$$\max_{x \in I} |E_n f(x)| \le \frac{\max\limits_{x \in I} |f^{(n+1)}(x)|}{4(n+1)} h^{n+1}. \qquad (3.7)$$

Unfortunately, we cannot deduce from (3.7) that the error tends to 0 when $n \to \infty$, in spite of the fact that $h^{n+1}/[4(n+1)]$ tends to 0. In fact, as shown in Example 3.2, there exist functions f for which the limit can even be infinite, that is

$$\lim_{n \to \infty} \max_{x \in I} |E_n f(x)| = \infty.$$

This striking result indicates that by increasing the degree n of the interpolating polynomial we do not necessarily obtain a better reconstruction of f. For instance, should we use all data of the first column of Table 3.1, we would obtain the interpolating polynomial $\Pi_{12} f$ represented in Figure 3.5, whose behavior in the vicinity of the left-hand of the interval is far less satisfactory than that obtained in Figure 3.4 using a much smaller number of nodes. An even worse result may arise for a special class of functions, as we report in the next Example.

Example 3.2 (Runge) If the function $f(x) = 1/(1 + x^2)$ is interpolated at equispaced nodes in the interval $I = (-5, 5)$, the error $\max_{x \in I} |E_n f(x)|$ tends to infinity when $n \to \infty$. This is due to the fact that if $n \to \infty$ the order of magnitude of $\max_{x \in I} |f^{(n+1)}(x)|$ outweighs the infinitesimal order of $h^{n+1}/[4(n+1)]$. This conclusion can be verified by computing the maximum of f and its derivatives up to the order 21 by means of the following instructions:

```
>> syms x; n=20; f='1/(1+x^2)'; df=diff(f,1);
>> cdf = char(df);
>> for i = 1:n+1, df = diff(df,1); cdfn = char(df);
   x = fzero(cdfn,0); M(i) = abs(eval(cdf)); cdf = cdfn;
   end
```

The absolute values of the functions $f^{(n)}$, $n = 1, \ldots, 21$, are stored in the vector M. Notice that the command **char** converts the symbolic expression **df** into a string that can be evaluated by the function **fzero**. In particular, the absolute values of $f^{(n)}$ for $n = 3, 9, 15, 21$ are:

```
>> M([3,9,15,21]) =
ans =
   4.6686e+00   3.2426e+05   1.2160e+12   4.8421e+19
```

while the corresponding values of the maximum of $\prod_{i=0}^{n}(x - x_i)/(n + 1)!$ are

```
>> z = linspace(-5,5,10000);
>> for n=0:20; h=10/(n+1); x=[-5:h:5];
   c=poly(x); r(n+1)=max(polyval(c,z));r(n+1)=r(n+1)/prod([1:n+2]); end
>> r([3,9,15,21])
ans =

   2.8935e+00   5.1813e-03   8.5854e-07   2.1461e-11
```

where c=poly(x) is a vector whose elements are the coefficients of the polyno- poly
mial whose roots are the elements of the vector **x**. It follows that $\max_{x \in I} |E_n f(x)|$
attains the following values:

```
>> format short e;
    1.3509e+01   1.6801e+03   1.0442e+06   1.0399e+09
```

for $n = 3, 9, 15, 21$, respectively.

The lack of convergence is also indicated by the presence of severe oscillations in the graph of the interpolating polynomial with respect to the graph
of f, especially near the endpoints of the interval (see Figure 3.5, right). This
behavior is known as *Runge's phenomenon*.

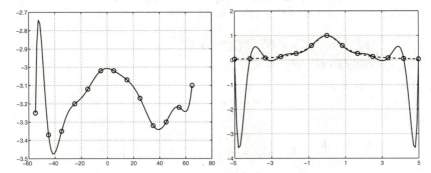

Fig. 3.5. Two examples of Runge's phenomenon: to the left, $\Pi_{12} f$ computed
for the data of Table 3.1, column $K = 0.67$; to the right, $\Pi_{12} f$ (solid line)
computed on 13 equispaced nodes for the function $f(x) = 1/(1 + x^2)$ (dashed
line)

Remark 3.1 The following inequality can also be proved:

$$\max_{x \in I} |f'(x) - (\Pi_n f)'(x)| \leq Ch^n \max_{x \in I} |f^{(n+1)}(x)|,$$

where C is a constant independent of h. Therefore, if we approximate the first
derivative of f by the first derivative of $\Pi_n f$, we loose an order of convergence
with respect to h. In MATLAB, $(\Pi_n f)'$ can be computed using the instruction
[d]=polyder(c), where c is the input vector in which we store the coefficients
of the interpolating polynomial, while d is the output vector where we store
the coefficients of its first derivative (see Section 1.4.2).

See the Exercises 3.1-3.4.

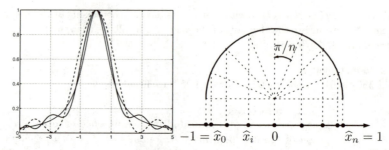

Fig. 3.6. The left side picture shows the comparison between the function $f(x) = 1/(1 + x^2)$ (thin solid line) and its Chebyshev interpolating polynomials of degree 8 (dashed line) and 12 (solid line). Note that the amplitude of spurious oscillations decreases as the degree increases. The right side picture shows the distribution of Chebyshev nodes in the interval $[-1, 1]$

3.1.2 Chebyshev interpolation

Runge's phenomenon can be avoided if a suitable distribution of nodes is used. In particular, in an arbitrary interval $[a, b]$, we can consider the so called *Chebyshev nodes* (see Figure 3.6, right):

$$x_i = \frac{a+b}{2} + \frac{b-a}{2}\widehat{x}_i, \text{ where } \widehat{x}_i = -\cos(\pi i/n), \quad i = 0, \ldots, n.$$

Indeed, for this special distribution of nodes it is possible to prove that, if f is a continuous and differentiable function in $[a, b]$, $\Pi_n f$ converges to f as $n \to \infty$ for all $x \in [a, b]$.

The Chebyshev nodes, which are the abscissas of equispaced nodes on the unit semi-circle, lie inside $[a, b]$ and are clustered near the endpoints of this interval (see Figure 3.6).

Another non-uniform distribution of nodes in the interval $[a, b]$, sharing the same convergence properties of Chebyshev nodes, is provided by:

$$\widehat{x}_i = \frac{a+b}{2} - \frac{b-a}{2}\cos\left(\frac{2i+1}{n+1}\frac{\pi}{2}\right), \quad i = 0, \ldots, n.$$

Example 3.3 We consider anew the function f of Runge's example and compute its interpolating polynomial at Chebyshev nodes. The latter can be obtained through the following MATLAB instructions:

```
>> xc = -cos(pi*[0:n]/n); x = (a+b)*0.5+(b-a)*xc*0.5;
```

where **n+1** is the number of nodes, while **a** and **b** are the endpoints of the interpolation interval (in the sequel we choose **a=-5** and **b=5**). Then we compute the interpolating polynomial by the following instructions:

```
>> f= '1./(1+x.^2)'; y = eval(f); c = polyfit(x,y,n);
```

Now let us compute the absolute values of the differences between f and its Chebyshev interpolant at as many as 1001 equispaced points in the interval $[-5, 5]$ and take the maximum error values:

```
>> x = linspace(-5,5,1000); p=polyval(c,x); fx = eval(f); err = max(abs(p-fx));
```

As we see in Table 3.2, the maximum of the error decreases when n increases.

n	5	10	20	40
E_n	0.6386	0.1322	0.0177	0.0003

Tab. 3.2. The Chebyshev interpolation error for Runge's function $f(x) = 1/(1 + x^2)$

3.1.3   Trigonometric interpolation and FFT

We want to approximate a periodic function $f : [0, 2\pi] \to \mathbb{R}$, i.e. one satisfying $f(0) = f(2\pi)$, by a trigonometric polynomial $\tilde{f}$ which interpolates f at the $n + 1$ nodes $x_j = 2\pi j/(n + 1)$, $j = 0, \ldots, n$, i.e.

$$\tilde{f}(x_j) = f(x_j), \text{ for } j = 0, \ldots, n. \tag{3.8}$$

The *trigonometric interpolant* $\tilde{f}$ is obtained by a linear combination of sines and cosines.

In particular, if n is even, $\tilde{f}$ will have the form

$$\tilde{f}(x) = \frac{a_0}{2} + \sum_{k=1}^{M} [a_k \cos(kx) + b_k \sin(kx)], \tag{3.9}$$

where $M = n/2$ while, if n is odd,

$$\tilde{f}(x) = \frac{a_0}{2} + \sum_{k=1}^{M} [a_k \cos(kx) + b_k \sin(kx)] + a_{M+1} \cos((M + 1)x), \tag{3.10}$$

where $M = (n - 1)/2$. We can rewrite (3.9) as

$$\tilde{f}(x) = \sum_{k=-M}^{M} c_k e^{ikx}, \tag{3.11}$$

i being the imaginary unit. The coefficients c_k are related to the coefficients a_k and b_k as follows:

$$a_k = c_k + c_{-k}, \quad b_k = i(c_k - c_{-k}), \quad k = 0, \dots, M. \tag{3.12}$$

Indeed, from (1.5) it follows that $e^{ikx} = \cos(kx) + i\sin(kx)$ and

$$\sum_{k=-M}^{M} c_k e^{ikx} = \sum_{k=-M}^{M} c_k \left(\cos(kx) + i\sin(kx) \right)$$

$$= \sum_{k=1}^{M} \left[c_k(\cos(kx) + i\sin(kx)) + c_{-k}(\cos(kx) - i\sin(kx)) \right] + c_0.$$

Therefore we derive (3.9), thanks to the relations (3.12).

Analogously, when n is odd, (3.10) becomes

$$\tilde{f}(x) = \sum_{k=-(M+1)}^{M+1} c_k e^{ikx}, \tag{3.13}$$

where the coefficients c_k for $k = 0, \dots, M$ are the same as before, while $c_{M+1} = c_{-(M+1)} = a_{M+1}/2$. In both cases, we could write

$$\tilde{f}(x) = \sum_{k=-(M+\mu)}^{M+\mu} c_k e^{ikx}, \tag{3.14}$$

with $\mu = 0$ if n is even and $\mu = 1$ if n is odd.

Because of its analogy with Fourier series, $\tilde{f}$ is called a *(real) discrete Fourier transform* (DFT). Imposing the interpolation condition at the nodes $x_j = jh$, with $h = 2\pi/(n+1)$, we find that

$$\sum_{k=-(M+\mu)}^{M+\mu} c_k e^{ikjh} = f(x_j), \quad j = 0, \dots, n. \tag{3.15}$$

For the computation of the coefficients $\{c_k\}$ let us multiply equations (3.15) by $e^{-imx_j} = e^{-imjh}$, where m is an integer between 0 and n, and then sum with respect to j:

$$\sum_{j=0}^{n} \sum_{k=-(M+\mu)}^{M+\mu} c_k e^{ikjh} e^{-imjh} = \sum_{j=0}^{n} f(x_j) e^{-imjh}. \tag{3.16}$$

We now require the following identity:

$$\sum_{j=0}^{n} e^{ijh(k-m)} = (n+1)\delta_{km},$$

where δ_{km} is the Kronecker symbol. This identity is obviously true if $k = m$. When $k \neq m$, we have

$$\sum_{j=0}^{n} e^{ijh(k-m)} = \frac{1 - \left(e^{i(k-m)h}\right)^{n+1}}{1 - e^{i(k-m)h}}.$$

The numerator on the right hand side is null, since

$$1 - e^{i(k-m)h(n+1)} = 1 - e^{i(k-m)2\pi} = 1 - \cos((k-m)2\pi) - i\sin((k-m)2\pi).$$

Therefore, from (3.16) we get the following explicit expression for the coefficients of $\tilde{f}$:

$$c_m = \frac{1}{n+1} \sum_{j=0}^{n} f(x_j) e^{-imjh}, \quad m = -(M+\mu), \ldots, M+\mu$$

The computation of all the coefficients c_m can be accomplished with an order $n \log_2 n$ operations by using the *fast Fourier transform* (FFT), which is implemented in the MATLAB program `fft` (see Example 3.4). `fft` Similar conclusions hold for the inverse transform through which we obtain the values $\{f(x_j)\}$ from the coefficients $\{c_k\}$. The inverse fast Fourier transform is implemented in the MATLAB program `ifft`.   `ifft`

Example 3.4 Consider the function $f(x) = x(x - 2\pi)e^{-x}$ for $x \in [0, 2\pi]$. To use the MATLAB program `fft`, we evaluate f at the nodes $x_j = j\pi/5$ for $j = 0, \ldots, 9$. Then, by the following instructions (and recalling that `.*` is the component-by-component vector product):

```
>> x=pi/5*[0:9]; y=x.*(x-2*pi).*exp(-x); Y=fft(y);
```

we compute

```
>> Y =
   Columns 1 through 3
   -6.5203e+00             -4.6728e-01 + 4.2001e+00i  1.2681e+00 + 1.6211e+00i
   Columns 4 through 6
   1.0985e+00 + 6.0080e-01i  9.2585e-01 + 2.1398e-01i  8.7010e-01 - 1.3887e-16i
   Columns 7 through 9
   9.2585e-01 - 2.1398e-01i  1.0985e+00 - 6.0080e-01i  1.2681e+00 - 1.6211e+00i
   Column 10
   -4.6728e-01 - 4.2001e+00i
```

The coefficients $\{c_k\}$ are simply given by `Y / length(Y)`, where the command `length` computes the dimension of a vector (10 in this example).

Note that the program `ifft` achieves the maximum efficiency when n is a power of 2, even though it works for any value of n.

The command `interpft` provides the trigonometric interpolant of a set of data. It requires in input an integer N and a vector of values which represent the values taken by a function (periodic with period p) at the set of points $x_i = ip/M$, $i = 1,\ldots,M-1$. `interpft` returns the N values of the trigonometric interpolant, obtained by the Fourier transform, at the nodes $t_i = ip/N$, $i = 0,\ldots,N-1$. For instance, let us reconsider the function of Example 3.4 in $[0, 2\pi]$ and take its values at 10 equispaced nodes $x_i = i\pi/5$, $i = 0,\ldots,9$. The values of the trigonometric interpolant at, say, the 100 equispaced nodes $t_i = i\pi/100$, $i = 0,\ldots,99$ can be obtained as follows (see Figure 3.7)

```
>>  x=pi/5*[0:9]; y=x.*(x-2*pi).*exp(-x); z=interpft(y,100);
```

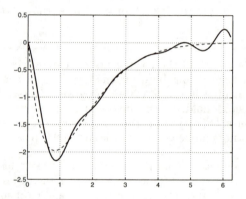

Fig. 3.7. The function $f(x) = x(x - 2\pi)e^{-x}$ (dashed line) and the corresponding trigonometric interpolant (continuous line) relative to 10 equispaced nodes

In some cases the accuracy of trigonometric interpolation can dramatically downgrade, as shown in the following example.

Example 3.5 Let us approximate the function $f(x) = f_1(x) + f_2(x)$ where $f_1(x) = \sin(x)$ and $f_2(x) = \sin(5x)$, using nine equispaced nodes in the interval $[0, 2\pi]$. The result is shown in Figure 3.8. Note that in some intervals the trigonometric approximant shows even a phase inversion with respect to the function f.

    This lack of accuracy can be explained as follows. At the nodes considered, the function f_2 is indistinguishable from $f_3(x) = -\sin(3x)$ which has a lower frequency (see Figure 3.9). The function that is actually approximated is therefore $F(x) = f_1(x) + f_3(x)$ and not $f(x)$ (in fact, the dashed line of Figure 3.8 does coincide with F).

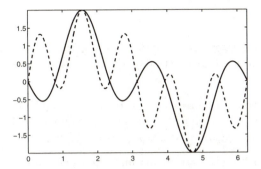

Fig. 3.8. The effects of aliasing: comparison between the function $f(x) = \sin(x) + \sin(5x)$ (solid line) and its trigonometric interpolant (3.9) with $M = 3$ (dashed line)

This phenomenon is known as *aliasing* and may occur when the function to be approximated is the sum of several components having different frequencies. As soon as the number of nodes is not enough to resolve the highest frequencies, the latter may interfere with the low frequencies, giving rise to inaccurate interpolants. To get a better approximation for functions with higher frequencies, one has to increase the number of interpolation nodes.

A real life example of aliasing is provided by the apparent inversion of the sense of rotation of spoked wheels. Once a certain critical velocity is reached the human brain is no longer able to accurately sample the moving image and, consequently, produces distorted images.

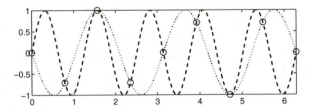

Fig. 3.9. The phenomenon of aliasing: the functions $\sin(5x)$ (dashed line) and $-\sin(3x)$ (dotted line) take the same values at the interpolation nodes. This circumstance explains the severe loss of accuracy shown in Figure 3.8

Let us summarize

1. Approximating a set of data or a function f in $[a, b]$ consists of finding a suitable function $\tilde{f}$ that represents them with enough accuracy;

2. the interpolation process consists of determining a function $\tilde{f}$ such that $\tilde{f}(x_i) = y_i$, where the $\{x_i\}$ are given nodes and $\{y_i\}$ are either the values $\{f(x_i)\}$ or a set of prescribed values;

3. if the $n+1$ nodes $\{x_i\}$ are distinct, there exists a unique polynomial of degree less than or equal to n interpolating a set of prescribed values $\{y_i\}$ at the nodes $\{x_i\}$;

4. for an equispaced distribution of nodes in $[a, b]$ the interpolation error at any point of $[a, b]$ does not necessarily tend to 0 as n tends to infinity. However, there exist special distributions of nodes, for instance the Chebyshev nodes, for which this convergence property holds true for all continuous functions;

5. trigonometric interpolation is well suited to approximate periodic functions, and is based on choosing $\tilde{f}$ as a linear combination of sine and cosine functions. The FFT is a very efficient algorithm which allows the computation of the Fourier coefficients of a trigonometric interpolant from its node values and admits an equally fast inverse, the IFFT.

3.2   Piecewise linear interpolation

The Chebyshev interpolant provides an accurate approximation of smooth functions f whose expression is known. In the case when f is non smooth or when f is only known by its values at a set of given points (which do not coincide with the Chebyshev nodes), one can resort to a different interpolation method which is called linear composite interpolation.

More precisely, given a set of nodes $x_0 < x_1 < \ldots < x_n$, we denote by I_i the interval $[x_i, x_{i+1}]$. We approximate f by a continuous function which, on each interval, is given by the segment joining the two points $(x_i, f(x_i))$ and $(x_{i+1}, f(x_{i+1}))$ (see Figure 3.10). This function, denoted by $\Pi_1^H f$, is called a *piecewise linear interpolation polynomial* and its expression is:

$$\Pi_1^H f(x) = f(x_i) + \frac{f(x_{i+1}) - f(x_i)}{x_{i+1} - x_i}(x - x_i) \quad \text{for } x \in I_i.$$

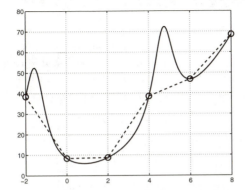

Fig. 3.10. The function $f(x) = x^2 + 10/(\sin(x) + 1.2)$ (solid line) and its piecewise linear interpolation polynomial $\Pi_1^H f$ (dashed line)

The upper-index H denotes the maximum length of the intervals I_i.

The following result can be inferred from (3.7) setting $n = 1$ and $h = H$:

Proposition 3.3 *If* $f \in C^2(I)$*, where* $I = [x_0, x_n]$*, then*

$$\max_{x \in I}|f(x) - \Pi_1^H f(x)| \leq \frac{H^2}{8}\max_{x \in I}|f''(x)|.$$

Consequently, for all x in the interpolation interval, $\Pi_1^H f(x)$ tends to $f(x)$ when $H \to 0$, provided that f is sufficiently smooth.

Through the instruction `s1=interp1(x,y,z)` one can compute the `interp1` values at arbitrary points, which are stored in the vector `z`, of the piecewise linear polynomial that interpolates the values `y(i)` at the nodes `x(i)`, for `i = 1,...,n+1`. Note that `z` can have arbitrary dimension. If the nodes are in increasing order (*i.e.* `x(i+1) > x(i)`, for `i=1,...,n`) then we can use the quicker version `interp1q` (q stands for quickly).   `interp1q`

It is worth mentioning that the command `fplot`, which is used to display the graph of a function f on a given interval $[a, b]$, does indeed replace the function by its piecewise linear interpolant. The set of interpolating nodes is generated automatically from the function, following the criterion of clustering these nodes around points where f shows strong variations. A procedure of this type is called *adaptive*.

3.3   Approximation by spline functions

The main drawback of piecewise linear interpolation is that $\Pi_1^H f$ is nothing more than a global continuous function. As a matter of fact, in several applications, e.g. in computer graphics, it is desirable to get approximation by smooth functions which have at least a continuous derivative.

With this aim, we can construct a function s_3 with the following properties:

1. on each interval $I_i = [x_i, x_{i+1}]$, for $i = 0, \ldots, n-1$ s_3 is a polynomial of degree 3 which interpolates the pairs of values $(x_j, f(x_j))$;

2. s_3 has continuous first and second derivatives on $[x_0, x_n]$.

For its complete determination, we need 4 conditions on each interval, therefore a total of $4n$ equations, which we can provide as follows: $n+1$ conditions arise from the interpolation requirement at the nodes x_j; requiring the continuity of the polynomial at the internal nodes $x_1, \ldots, x_{n-1}$ yields $n-1$ further equations; we obtain $2(n-1)$ new equations by requiring that both first and second derivatives be continuous at the internal nodes. Finally, we still lack two further equations, which we can e.g. choose as

$$s_3''(x_0) = 0, \quad s_3''(x_n) = 0. \tag{3.17}$$

The function s_3 which we obtain in this way, is called a *natural interpolating cubic spline*. By choosing suitably the unknowns (see [QSS00], Section 8.6.1) to represent s_3 we arrive at a $(n+1) \times (n+1)$ system with a tridiagonal matrix whose solution can be accomplished by a number of operations proportional to n (see Section 5.4).

The choice (3.17) is not the only one possible to complete the system of equations. Several alternatives do exist. One possibility would be to replace (3.17) by requiring the continuity of the third derivative of s_3 at the nodes x_1 and x_{n-1}. This is precisely what we get when using the MATLAB command spline (see also the toolbox splines). The input parameters are the vectors x and y, as well as the vector z containing the points at which we are seeking the values of s_3.

spline

Example 3.6 Let us reconsider the data of Table 3.1 corresponding to the column $K = 0.67$ and compute the associated cubic spline s_3. The different values of the latitude provide the nodes x_i, $i = 0, \ldots, 12$. If we are interested in computing the values $s_3(z_i)$, where $z_i = -55 + i$, $i = 0, \ldots, 120$, we can proceed as follows:

```
>> x = [-55:10:65];
>> y = [-3.25  -3.37  -3.35  -3.2  -3.12  -3.02  -3.02 ...
   -3.07  -3.17  -3.32  -3.3  -3.22  -3.1];
>> z = [-55:1:65];
>> s = spline(x,y,z);
```

The graph of s_3, which is reported in Figure 3.11, looks more plausible than that of the Lagrange interpolant at the same nodes.

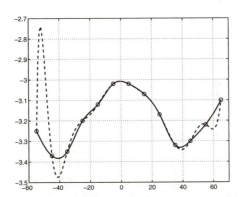

Fig. 3.11. Comparison between the cubic spline and the Lagrange interpolant for the case considered in Example 3.6

The error that we obtain in approximating a function f (continuously differentiable up to its fourth derivative) by the natural cubic spline satisfies the following inequalities:

$$\max_{x\in I}|f^{(r)}(x) - s_3^{(r)}(x)| \le C_r H^{4-r}\max_{x\in I}|f^{(4)}(x)|, \quad r = 0,1,2,3,$$

where $I = [x_0, x_n]$ and $H = \max_{i=0,\dots,n-1}(x_{i+1} - x_i)$, while C_r is a suitable constant depending on r, but independent of H. It is then clear that not only f, but also its first, second and third derivatives are well approximated by s_3 when H tends to 0.

Remark 3.2 Cubic splines in general don't preserve monotonicity between neighbouring nodes. For instance, by approximating the unitary circumference in the first quarter using the points $(x_k = \sin(k\pi/6), y_k = \cos(k\pi/6))$, for $k = 0,\dots,3$, we would obtain an oscillatory spline (see Figure 3.12). In these cases, other approximation techniques can be better suited. For instance, the MATLAB command `pchip` provides the Hermite piecewise cubic interpolant   pchip and guarantees the local monotonicity of the interpolant (see Figure 3.12). The Hermite interpolant can be obtained by using the following instructions:

```
>> t = linspace(0,pi/2,4)
>> x = cos(t); y = sin(t);
>> xx = linspace(0,1,40);
>> plot(x,y,'s',xx,[pchip(x,y,xx);spline(x,y,xx)])
```

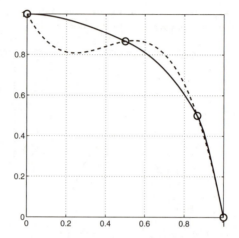

Fig. 3.12. Approximation of the first quarter of the circumference of the unitary circle using only 4 nodes. The dashed line is the cubic spline, whereas the continuous line is the piecewise cubic Hermite interpolant

See the Exercises 3.5-3.8.

3.4   The least squares method

We have already noticed that a Lagrange interpolation does not guarantee a better approximation of a given function when the polynomial degree gets large. This problem can be overcome by composite interpolation (such as piecewise linear polynomials or splines). However, neither are suitable to extrapolate information from the available data, that is, to generate new values at points lying outside the interval where interpolation nodes are given.

Example 3.7 On the basis of the data reported in Figure 3.1, we would like to predict whether the stock price will increase or diminish in the coming days. The Lagrange polynomial interpolation is impractical, as it would require a (tremendously oscillatory) polynomial of degree 719 which will provide a completely useless prediction. On the other hand, piecewise linear interpolation, whose graph is reported in Figure 3.1, provides extrapolated results by exploiting only the values of the last two days, thus completely neglecting the previous history. To get a better result we should avoid the interpolation requirement, by invoking least square approximation as indicated below.

Assume that the data $\{(x_i, f(x_i)), i = 0, \ldots, n\}$ are available. We look for a polynomial $\tilde{f}$ of degree at most $m \geq 1$ which satisfies the following inequality:

$$\sum_{i=0}^{n} [f(x_i) - \tilde{f}(x_i)]^2 \leq \sum_{i=0}^{n} [f(x_i) - p_m(x_i)]^2 \qquad (3.18)$$

for every polynomial p_m of degree at most m. Should it exist, $\tilde{f}$ will be called the *least square approximation* of f. For arbitrary values of m and n, it will not be possible to guarantee that $f(x_i) = \tilde{f}(x_i)$ for all $i = 0, \ldots, n$.

Setting

$$\tilde{f}(x) = a_0 + a_1 x + \ldots + a_m x^m, \qquad (3.19)$$

where the coefficients $a_0, \ldots, a_m$ are unknown, the problem (3.18) can be restated as follows:

$$\Phi(a_0, a_1, \ldots, a_m) = \min_{\{b_i, \; i=0,\ldots,m\}} \Phi(b_0, b_1, \ldots, b_m)$$

where

$$\Phi(b_0, b_1, \ldots, b_m) = \sum_{i=0}^{n} [f(x_i) - (b_0 + b_1 x_i + \ldots + b_m x_i^m)]^2.$$

Let us solve this problem in the special case when $m = 1$. Since

$$\Phi(b_0, b_1) = \sum_{i=0}^{n} \left[f(x_i)^2 + b_0^2 + b_1^2 x_i^2 + 2b_0 b_1 x_i - 2b_0 f(x_i) - 2b_1 x_i f(x_i) \right],$$

the graph of Φ is a convex paraboloid. The point (a_0, a_1) at which Φ attains its minimum satisfies the conditions:

$$\frac{\partial \Phi}{\partial b_0}(a_0, a_1) = 0, \quad \frac{\partial \Phi}{\partial b_1}(a_0, a_1) = 0,$$

where the symbol $\partial \Phi / \partial b_j$ denotes the partial derivative (that is, the rate of variation) of Φ with respect to b_j, after having frozen the remaining variable (see the definition 8.3).

By explicitly computing the two partial derivatives we obtain

$$\sum_{i=0}^{n}[a_0 + a_1 x_i - f(x_i)] = 0, \quad \sum_{i=0}^{n}[a_0 x_i + a_1 x_i^2 - x_i f(x_i)] = 0,$$

which is a system of 2 equations for the 2 unknowns a_0 and a_1:

$$
\begin{aligned}
a_0(n+1) + a_1\sum_{i=0}^{n} x_i &= \sum_{i=0}^{n} f(x_i), \\
a_0\sum_{i=0}^{n} x_i + a_1\sum_{i=0}^{n} x_i^2 &= \sum_{i=0}^{n} f(x_i)x_i.
\end{aligned}
\tag{3.20}
$$

Setting $D = (n+1)\sum_{i=0}^{n} x_i^2 - (\sum_{i=0}^{n} x_i)^2$, the solution reads:

$$
\begin{aligned}
a_0 &= \frac{1}{D}\left[\sum_{i=0}^{n} f(x_i)\sum_{j=0}^{n} x_j^2 - \sum_{j=0}^{n} x_j\sum_{i=0}^{n} x_i f(x_i)\right], \\
a_1 &= \frac{1}{D}\left[(n+1)\sum_{i=0}^{n} x_i f(x_i) - \sum_{j=0}^{n} x_j\sum_{i=0}^{n} f(x_i)\right].
\end{aligned}
\tag{3.21}
$$

The corresponding polynomial $\tilde{f}$ is known as the *least square straight line*, or *regression line*.

The previous approach can be generalized to arbitrary m. The associated $(m+1) \times (m+1)$ linear system, which is symmetric, will have the following form:

$$
\begin{aligned}
a_0(n+1) &+a_1\sum_{i=0}^{n} x_i &+\ldots+ a_m\sum_{i=0}^{n} x_i^m &= \sum_{i=0}^{n} f(x_i), \\
a_0\sum_{i=0}^{n} x_i &+a_1\sum_{i=0}^{n} x_i^2 &+\ldots+ a_m\sum_{i=0}^{n} x_i^{m+1} &= \sum_{i=0}^{n} x_i f(x_i), \\
&\vdots \quad\quad\quad &\vdots \quad\quad\quad &\vdots \\
a_0\sum_{i=0}^{n} x_i^m &+a_1\sum_{i=0}^{n} x_i^{m+1} &+\ldots+ a_m\sum_{i=0}^{n} x_i^{2m} &= \sum_{i=0}^{n} x_i^m f(x_i).
\end{aligned}
$$

When $m = n$, the least square polynomial must coincide with the Lagrange interpolating polynomial $\Pi_n f$ (see Exercise 3.9).

The MATLAB command c=polyfit(x,y,m) computes by default the coefficients of the polynomial of degree m which approximates n+1 pairs of data (x(i),y(i)) in the least square sense. As already noticed in Section 3.1.1, when m is equal to n it returns the interpolating polynomial.

Example 3.8 In Figure 3.13 we draw the graphs of the least square polynomials of degree 1, 2 and 4 that approximate the data of Figure 3.1. The polynomial of degree 4 reproduces quite reasonably the behavior of the stock price in the considered time interval and suggests that in the near future the quotation will increase.

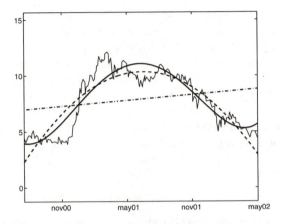

Fig. 3.13. Least squares approximation of the data of Problem 3.2 of degree 1 (dashed-dotted line), degree 2 (dashed line) and degree 4 (thick solid line). The exact data are represented by the thin solid line

Let us summarize

1. The composite piecewise linear interpolant of a function f is a piecewise continuous linear function $\tilde{f}$, which interpolates f at a given set of nodes $\{x_i\}$. With this approximation we avoid Runge's type phenomena when the number of nodes increases;

2. interpolation by cubic splines allows the approximation of f by a piecewise cubic function $\tilde{f}$ which is continuous together with its first and second derivatives;

3. in least squares approximation we look for an approximant $\tilde{f}$ which is a polynomial of degree m (typically, $m < n$) that minimizes the mean-square error $\sum_{i=0}^{n} [f(x_i) - \tilde{f}(x_i)]^2$.

See the Exercises 3.9-3.13.

3.5   What we haven't told you

For a more general introduction to the theory of interpolation and approximation the reader is referred to [Dav63], [Mei67] and [Gau97].

Polynomial interpolation can also be used to approximate data and functions in several dimensions. In particular, composite interpolation, based on piecewise linear or spline functions, is well suited when the region Ω at hand is partitioned into polygons in 2D (triangles or quadrilaterals) and polyhedra in 3D (tetrahedra or prisms).

A special situation occurs when Ω is a rectangle or a parallelepiped. In
interp2   that case one could simply use the commands zi=interp2(x,y,z,xi,
interp3   yi) or vi=interp3(x,y,z,v,xi,yi,zi), respectively, x, ..., v being the interpolating values and xi, ..., zi the nodes at which the interpolating polynomial should be evaluated.

For instance, to approximate by a cubic spline the values of the function $f(x,y) = \sin(2\pi x)\cos(2\pi y)$ on a uniform grid of 36 nodes on the square $[0,1]^2$, we use the following instructions:

```
>> [x,y]=meshgrid(0:0.2:1,0:0.2:1); z=sin(2*pi*x).*cos(2*pi*y);
```

Then, the cubic interpolating spline, evaluated on a uniform grid of 441 nodes (21 in both x and y directions), can be obtained as follows:

```
>> xi = [0:0.05:1]; yi=[0:0.05:1];
>> [xf,yf]=meshgrid(0:0.05:1,0:0.05:1); pi3=interp2(x,y,z,xf,yf,'spline');
```

meshgrid   The command meshgrid transforms the domain specified by the vectors x and y into arrays xf and yf that can be used for the evaluation of a function of two variables and for 3-dimensional MATLAB surface plots. The rows of the matrix xf are copies of the vector xi and the columns of the matrix yf are copies of the vector yi. Alternatively, one can use the
griddata   function griddata (also available for 3-dimensional data (griddata3) or for n-dimensional hypersurface fitting (griddatan).

When Ω is a two-dimensional domain of arbitrary shape, it can be
pdetool   partitioned into triangles using the graphical interface pdetool.

For a general presentation of spline functions see, e.g., [Die93] and [PBP02]. The MATLAB toolbox splines allows one to explore several
spdemos   applications of spline functions. In particular, the spdemos command gives the user the possibility to investigate the properties of the most important type of spline functions. Rational splines, i.e. functions which are the ratio of two splines functions, are accessible through the com-
rpmak   mands rpmak and rsmak. Special instances are the so-called NURBS
rsmak   splines, which are commonly used in CAGD (*Computer Assisted Geometric Design*).

In the same context of Fourier approximation, we mention the approximation based on *wavelets*. This type of approximation is largely used for image reconstruction and compression and in signal analysis (for an introduction, see [DL92], [Urb02]). A rich family of wavelets (and their applications) can be found in the MATLAB toolbox `wavelet`.

`wavelet`

3.6   Exercises

Exercise 3.1 Prove inequality (3.6).

Exercise 3.2 Provide an upper bound of the Lagrange interpolation error for the following functions:

$$f_1(x) = \cosh(x), \ f_2(x) = \sinh(x), \quad x_k = -1 + 0.5k, \ k = 0, \ldots, 4,$$
$$f_3(x) = \cos(x) + \sin(x), \qquad\qquad x_k = -\pi/2 + \pi k/4, \ k = 0, \ldots, 4.$$

Exercise 3.3 The following data are related to the life expectation of citizens of two European regions:

	1975	1980	1985	1990
Western Europe	72.8	74.2	75.2	76.4
Eastern Europe	70.2	70.2	70.3	71.2

Use the interpolating polynomial of degree 3 to estimate the life expectation in 1970, 1983 and 1988. Then extrapolate a value for the year 1995. It is known that the life expectation in 1970 was 71.8 years for the citizens of the West Europe, and 69.6 for those of the East Europe. Recalling these data, is it possible to estimate the accuracy of life expectation predicted in the 1995?

Exercise 3.4 The price (in euros) of a magazine has changed as follows:

Nov.87	Dec.88	Nov.90	Jan.93	Jan.95	Jan.96	Nov.96	Nov.00
4.5	5.0	6.0	6.5	7.0	7.5	8.0	8.0

Estimate the price in November 2002 by extrapolating these data.

Exercise 3.5 Repeat the computations carried out in Exercise 3.3, using now the cubic interpolating spline computed by the function `spline`. Then compare the results obtained with the two approaches.

Exercise 3.6 In the table below we report the values of the sea water density ρ (in Kg/m^3) corresponding to different values of the temperature T (in degrees Celsius):

T	4°	8°	12°	16°	20°
ρ	1000.7794	1000.6427	1000.2805	999.7165	998.9700

Compute the associated cubic interpolating spline on 4 subintervals of the temperature interval $[4, 20]$. Then compare the results provided by the spline interpolant with the following ones (which correspond to further values of T):

T	6°	10°	14°	18°
ρ	1000.74088	1000.4882	1000.0224	999.3650

Exercise 3.7 The Italian production of citrus fruit has changed as follows:

year	1965	1970	1980	1985	1990	1991
production ($\times 10^5$ Kg)	17769	24001	25961	34336	29036	33417

Use a cubic interpolating spline to estimate the production in 1962, 1977 and 1992. Compare these results with the real values: 12380, 27403 and 32059, respectively. Repeat the computation using the Lagrange interpolation polynomial.

Exercise 3.8 Evaluate the function $f(x) = \sin(2\pi x)$ at 21 equispaced nodes in the interval $[-1, 1]$. Compute the Lagrange interpolating polynomial and the cubic interpolating spline. Compare the graphs of these two functions with that of f on the given interval. Repeat the same calculation using the following perturbed set of data: $f(x_i) = (-1)^{i+1} 10^{-4}$, and observe that the Lagrange interpolating polynomial is more sensitive to small perturbations than the cubic spline.

Exercise 3.9 Verify that if $m = n$ the least-square polynomial of a function f at the nodes $x_0, \ldots, x_n$ coincides with the interpolating polynomial $\Pi_n f$ at the same nodes.

Exercise 3.10 Compute the least-squares polynomial of degree 4 that approximates the data of the second, third and fourth columns of Table 3.1.

Exercise 3.11 Repeat the computations carried out in Exercise 3.7 using now a least-squares approximation of degree 3.

Exercise 3.12 Express the coefficients of system (3.20) in terms of the *average* $M = \frac{1}{(N+1)} \sum_{i=0}^{n} x_i$ and the *variance* $v = \frac{1}{(N+1)} \sum_{i=0}^{n} (x_i - M)^2$ of the set of data $\{x_i, i = 0, \ldots, n\}$.

Exercise 3.13 Verify that the regression line passes through the point whose abscissa is the average of $\{x_i\}$ and ordinate is the average of $\{f(x_i)\}$.

4. Numerical differentiation and integration

In this chapter we propose methods for the numerical approximation of derivatives and integrals of functions. Concerning integration, it is known that for a generic function it is not always possible to find a primitive in an explicit form. In other cases, it could be hard to evaluate a primitive as, for instance, in the example

$$\frac{1}{\pi} \int_0^\pi \cos(4x) \cos(3 \sin(x)) \; dx = \left(\frac{3}{2}\right)^4 \sum_{k=0}^\infty \frac{(-9/4)^k}{k!(k+4)!},$$

where the problem of computing an integral is transformed into the equally troublesome one of summing a series. It is also worth mentioning that sometimes the function that we want to integrate or differentiate could only be known on a set of nodes (for instance, when the latter represent the results of an experimental measurement), exactly as happens in the case of the approximation of functions, which was discussed in chapter 3.

In all these situations it is necessary to consider numerical methods in order to obtain an approximate value of the quantity of interest, independently of how difficult is the function to integrate or differentiate.

Problem 4.1 (Hydraulics) The height $q(t)$ (in meters) of a flow in a conical funnel with a circular hole of radius $r = 10$ cm is measured at different time intervals, yielding the following values:

t	0.0	5	10	15	20
$q(t)$	1.0	0.8811	0.7366	0.5430	0.1698

We want to compute an approximation of the emptying velocity $q'(t)$ and compare it with the analytical one, given by $q'(t) = -0.6\pi r^2 \sqrt{19.6q(t)}/A(t)$

($A(t)$ is the area of the horizontal section of the cone corresponding to a fluid height $q(t)$). For the solution of this problem, see Example 4.1.   ●

Problem 4.2 (Optics) In order to plan a room for infrared beams we are interested in calculating the energy emitted by a black body (that is, an object capable of irradiating in all the spectrum to the ambient temperature) in the (infrared) spectrum comprised between 3μm and 14μm wavelength. The solution of this problem is obtained by computing the integral

$$E(T) = 2.39 \cdot 10^{-11} \int_{3\cdot 10^{-4}}^{14\cdot 10^{-4}} \frac{dx}{x^5(e^{1.432/(Tx)} - 1)}, \tag{4.1}$$

which is the Planck equation for the energy $E(T)$, where x is the wavelength (in cm) and T the temperature (in Kelvin) of the black body. For the computation of the above integral, see Exercise 4.17.   ●

Problem 4.3 (Electromagnetism) Consider an electric wire sphere of arbitrary radius r and conductivity σ. We want to compute the density distribution of the current $\mathbf{j}$ as a function of r and t (the time), knowing the initial distribution of the current density $\rho(r)$. The problem can be solved using the relations between the current density, the electric field and the charge density and observing that, for the symmetry of the problem, $\mathbf{j}(r,t) = j(r,t)\mathbf{r}/|\mathbf{r}|$, where $j = |\mathbf{j}|$. We obtain

$$j(r,t) = j(r)e^{-\sigma t/\varepsilon_0}, \quad j(r) = \frac{\sigma}{\varepsilon_0 r^2} \int_0^r \rho(\xi)\xi^2 \, d\xi, \tag{4.2}$$

where $\varepsilon_0 = 8.859 \cdot 10^{-12}$ farad/m is the dielectric constant of the void.
For the computation of $j(r)$, see Exercise 4.16.   ●

4.1 Approximation of function derivatives

Consider a function $f : [a, b] \to \mathbb{R}$ continuously differentiable in $[a, b]$. We seek an approximation of the first derivative of f at a generic point $\bar{x}$ in (a, b).

In view of the definition (1.9), for h sufficiently small and positive, we can assume that the quantity

$$(\delta_+ f)(\bar{x}) = \frac{f(\bar{x} + h) - f(\bar{x})}{h} \tag{4.3}$$

is an approximation of $f'(\bar{x})$ which is called the *forward finite difference*. To estimate the error, it suffices to expand f in a Taylor series, obtaining

$$f(\bar{x} + h) = f(\bar{x}) + hf'(\bar{x}) + \frac{h^2}{2}f''(\xi) \qquad (4.4)$$

where ξ is a suitable point in the interval $(\bar{x}, \bar{x} + h)$. Therefore

$$(\delta_+ f)(\bar{x}) = f'(\bar{x}) + \frac{h}{2}f''(\xi), \qquad (4.5)$$

and thus $(\delta_+ f)(\bar{x})$ provides a first-order approximation to $f'(\bar{x})$ with respect to h. With a similar procedure, we can derive from the Taylor expansion

$$f(\bar{x} - h) = f(\bar{x}) - hf'(\bar{x}) + \frac{h^2}{2}f''(\eta) \qquad (4.6)$$

with $\eta \in (\bar{x} - h, \bar{x})$, the *backward finite difference*

$$\boxed{(\delta_- f)(\bar{x}) = \frac{f(\bar{x}) - f(\bar{x} - h)}{h}} \qquad (4.7)$$

which is also first-order accurate. Note that formulae (4.3) and (4.7) can also be obtained by differentiating the linear polynomial interpolating f at the points $\{\bar{x}, \bar{x} + h\}$ and $\{\bar{x} - h, \bar{x}\}$, respectively. In fact, from the geometrical viewpoint, these schemes amount to approximating $f'(\bar{x})$ by the slope of the straight line passing through the two points $(\bar{x}, f(\bar{x}))$ and $(\bar{x} + h, f(\bar{x} + h))$, or $(\bar{x} - h, f(\bar{x} - h))$ and $(\bar{x}, f(\bar{x}))$, respectively (see Figure 4.1).

Finally, we introduce the *centered finite difference* formula

$$\boxed{(\delta f)(\bar{x}) = \frac{f(\bar{x} + h) - f(\bar{x} - h)}{2h}} \qquad (4.8)$$

which provides a second-order approximation to $f'(\bar{x})$ with respect to h. Indeed, we obtain

$$f'(\bar{x}) - (\delta f)(\bar{x}) = \frac{h^2}{12}[f'''(\xi) + f'''(\eta)], \qquad (4.9)$$

where η and ξ are suitable points in the intervals $(\bar{x} - h, \bar{x})$ and $(\bar{x}, \bar{x} + h)$, respectively (see Exercise 4.2).

By (4.8) $f'(\bar{x})$ is approximated by the slope of the straight line passing through the points $(\bar{x} - h, f(\bar{x} - h))$ and $(\bar{x} + h, f(\bar{x} + h))$.

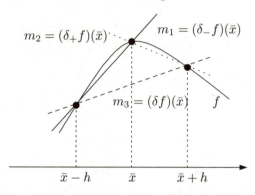

Fig. 4.1. Finite difference approximation of $f'(\bar{x})$: backward (solid line), forward (dotted line) and centered (dashed line). m_1, m_2 and m_3 denote the slopes of the three straight lines

Example 4.1 Let us solve Problem 4.1, using formulae (4.3), (4.7) and (4.8) to approximate $q'(t)$. We obtain:

t	0.0	5	10	15	20
$q'(t)$	-0.0220	-0.0259	-0.0326	-0.0473	-0.1504
δ_+q	-0.0238	-0.0289	-0.0387	-0.0746	$--$
δ_-q	$--$	-0.0238	-0.0289	-0.0387	-0.0746
δq	$--$	-0.0263	-0.0338	-0.0567	$--$

The agreement between the exact derivative and the one computed from the finite difference formulae with $h = 5$ is more satisfactory when using formula (4.8) rather than (4.7) or (4.3).

In general, we can assume that the values of f are available at $n + 1$ equispaced points $x_i = x_0 + ih$, with $h > 0$. In this case in the numerical derivation $f'(x_i)$ can be approximated by taking one of the previous formulae (4.3), (4.7) or (4.8) with $\bar{x} = x_i$.

Note that the centered formula (4.8) can be used only at the internal points $x_1, \ldots, x_{n-1}$ but not at the extrema x_0 and x_n. At the latter nodes we could use the values

$$\frac{1}{2h}\left[-3f(x_0) + 4f(x_1) - f(x_2)\right] \qquad \text{at } x_0,$$

$$\frac{1}{2h}\left[3f(x_n) - 4f(x_{n-1}) + f(x_{n-2})\right] \quad \text{at } x_n,$$

(4.10)

which are also second-order accurate with respect to h. They are obtained by computing at the point x_0 (respectively, x_n) the first derivative of the polynomial of degree 2 interpolating f at the nodes x_0, x_1, x_2 (respectively, x_{n-2}, x_{n-1}, x_n).

See the Exercises 4.1-4.4.

4.2   Numerical integration

In this section we introduce numerical methods suitable for approximating the integral

$$I(f) = \int_a^b f(x) \, dx,$$

where f is an arbitrary continuous function in $[a, b]$. We start by introducing some simple formulae, which are indeed special instances of the broader family of Newton-Cotes formulae.

4.2.1   Midpoint formula

A simple procedure to approximate $I(f)$ can be devised by partitioning the interval $[a, b]$ into subintervals $I_k = [x_{k-1}, x_k]$, $k = 1, \ldots, M$, with $x_k = a + kH$, $k = 0, \ldots, M$ and $H = (b-a)/M$. Since

$$I(f) = \sum_{k=1}^M \int_{I_k} f(x) \, dx, \qquad (4.11)$$

on each sub-interval I_k we can approximate the exact integral of f by that of a polynomial $\bar{f}$ approximating f on I_k. The simplest solution consists of choosing $\bar{f}$ as the constant polynomial interpolating f at the middle point of I_k:

$$\bar{x}_k = \frac{x_{k-1} + x_k}{2}.$$

In such a way we obtain the *composite midpoint quadrature formula*

$$I_{mp}^c(f) = H \sum_{k=1}^M f(\bar{x}_k) \qquad (4.12)$$

The symbol mp stands for midpoint, while c stands for composite. This formula is second-order accurate with respect to h. More precisely, if f is continuously differentiable up to its second derivative, we have

$$I(f) - I_{mp}^c(f) = \frac{b-a}{24} H^2 f''(\xi), \qquad (4.13)$$

where ξ is a suitable point in (a, b) (see Exercise 4.6). Formula (4.12) is also called the *composite rectangle quadrature formula* because of its geometrical interpretation, which is evident from Figure 4.2.

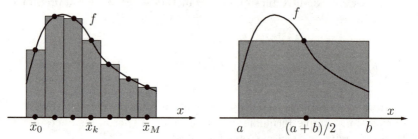

Fig. 4.2. The composite midpoint formula (left); the midpoint formula (right)

The classical *midpoint formula* (or *rectangle formula*) is obtained by taking $M = 1$ in (4.12), *i.e.* using the midpoint rule directly on the interval (a, b):

$$I_{mp}(f) = (b - a)f[(a + b)/2]$$

The error is now given by

$$I(f) - I_{mp}(f) = \frac{(b - a)^3}{24}f''(\xi) \tag{4.14}$$

where ξ is a suitable point in (a, b). Relation (4.14) follows as a special case of (4.13), but it can also be proved directly. Indeed, setting $\bar{x} = (a + b)/2$, we have

$$
\begin{aligned}
I(f) - I_{mp}(f) &= \int_a^b [f(x) - f(\bar{x})]\, dx \\
&= \int_a^b f'(\bar{x})(x - \bar{x})\, dx + \frac{1}{2}\int_a^b f''(\eta(x))(x - \bar{x})^2\, dx,
\end{aligned}
$$

where $\eta(x)$ is a suitable point in the interval whose endpoints are x and $\bar{x}$. Then (4.14) follows because $\int_a^b (x - \bar{x})\, dx = 0$ and, by the mean value theorem, $\exists \xi \in (a, b)$ such that

$$\frac{1}{2}\int_a^b f''(\eta(x))(x - \bar{x})^2\, dx = \frac{1}{2}f''(\xi)\int_a^b (x - \bar{x})^2\, dx = \frac{(b - a)^3}{24}f''(\xi).$$

The *degree of exactness* of a quadrature formula is the maximum integer $r \geq 0$ for which the approximate integral (produced by the quadrature formula) of any polynomial of degree r is equal to the exact integral. Thus, the midpoint formula has degree of exactness 1, since it integrates exactly all polynomials of degree less than or equal to 1 (but not all those of degree 2).

The midpoint composite quadrature formula is implemented in Program 3. Input parameters are the endpoints of the integration interval a and b, the number of subintervals M and a string f to define the function f. In particular, in order to ensure a correct execution of the program if f is a constant function, f must be provided in the form 'c+0.*x', where c is the constant value of f.

Program 3 - midpointc: composite midpoint quadrature formula

```
function Imp=midpointc(a,b,M,f)
%MIDPOINTC Composite midpoint numerical integration.
%   IMP = MIDPOINTC(A,B,M,FUN) computes an approximation of the integral
%   of the function FUN via the midpoint method (with M equispaced intervals).
%   FUN accepts real scalar input x and returns a real scalar value.
%   FUN can also be an inline object.
H=(b-a)/M;
x = linspace(a+H/2,b-H/2,M);
fmp=feval(f,x);
Imp=H*sum(fmp);
```

See the Exercises 4.5-4.8.

4.2.2 Trapezoidal formula

Another formula can be obtained by replacing f on I_k by the linear polynomial interpolating f at the nodes x_{k-1} and x_k (equivalently, replacing f by $\Pi_1^H f$, see Section 3.2, on the whole interval (a, b)). This yields

$$
\begin{aligned}
I_t^c(f) &= \frac{H}{2} \sum_{k=1}^{M} [f(x_k) + f(x_{k-1})] \\
&= \frac{H}{2} [f(a) + f(b)] + H \sum_{k=1}^{M-1} f(x_k)
\end{aligned}
\tag{4.15}
$$

This formula is called the *composite trapezoidal formula*, and is second-order accurate with respect to H.

In fact, one can obtain the expression

$$I(f) - I_t^c(f) = -\frac{b-a}{12}H^2 f''(\xi)$$ (4.16)

for the quadrature error for a suitable point $\xi \in (a, b)$, provided that $f \in C^2(a, b)$. When (4.15) is used with $M = 1$, we obtain

$$I_t(f) = \frac{b-a}{2}[f(a) + f(b)]$$ (4.17)

which is called the *trapezoidal formula* because of its geometrical interpretation. The error induced is given by

$$I(f) - I_t(f) = -\frac{(b-a)^3}{12}f''(\xi),$$ (4.18)

where ξ is a suitable point in (a, b), from which we can deduce that (4.17) has degree of exactness equal to 1, as is the case of the midpoint rule.

With a simple modification of this procedure we can obtain a more accurate formula. In fact, we can still approximate f by a polynomial, but this time we consider as interpolatory points the *Gauss nodes*

$$\gamma_{k-1} = \frac{x_{k-1} + x_k}{2} - \frac{1}{\sqrt{3}}\left(\frac{x_k - x_{k-1}}{2}\right) = x_{k-1} + \left(1 - \frac{1}{\sqrt{3}}\right)\frac{H}{2},$$

$$\gamma_k = \frac{x_{k-1} + x_k}{2} + \frac{1}{\sqrt{3}}\left(\frac{x_k - x_{k-1}}{2}\right) = x_{k-1} + \left(1 + \frac{1}{\sqrt{3}}\right)\frac{H}{2}.$$

With this choice we obtain the *Gauss quadrature formula*

$$I_{Gauss}^c(f) = \frac{H}{2}\sum_{k=0}^{M}[f(\gamma_{k-1}) + f(\gamma_k)]$$ (4.19)

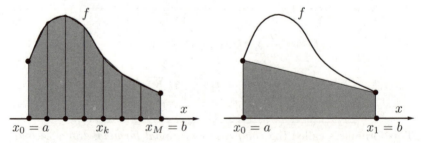

Fig. 4.3. Composite trapezoidal formula (left); trapezoidal formula (right)

Its order of accuracy (with respect to H) is equal to 4. Precisely,

$$I(f) - I^c_{Gauss}(f) = \frac{b-a}{4320} H^4 f^{(4)}(\xi)$$

where ξ is a suitable point in (a, b) (see [RR85]). Restricted to only one interval, (4.19) becomes:

$$I_{Gauss}(f) = \frac{(b-a)}{2} [f(\gamma_1) + f(\gamma_0)] \qquad (4.20)$$

where

$$\gamma_0 = \frac{a+b}{2} - \frac{1}{\sqrt{3}} \left(\frac{b-a}{2}\right), \quad \gamma_1 = \frac{a+b}{2} + \frac{1}{\sqrt{3}} \left(\frac{b-a}{2}\right).$$

The corresponding error is

$$I(f) - I_{Gauss}(f) = \frac{(b-a)^5}{4320} f^{(4)}(\eta),$$

for a suitable $\eta \in (a, b)$. In particular, it follows that this formula has degree of exactness equal to 3, i.e. it computes exactly the integral of all polynomials of degree less than or equal to 3 (but not those of degree 4). The quadrature formulae (4.15) and (4.19) are implemented in Program 4. The selection criterion is to set choice=1 for the trapezoidal rule, choice=2 for the Gauss rule. The trapezoidal formula (4.17) is implemented in the MATLAB program trapz, the composite formula (4.15) trapz in cumtrapz.   cumtrapz

Program 4 - trapc: composite trapezoidal and Gauss formulae

```
function [Itpc]=trapc(a,b,M,f,choice,varargin)
%TRAPC Composite two-points numerical integration.
%   ITPC = TRAPC(A,B,M,FUN,CHOICE) computes an approximation of the
%   integral of the function FUN via the trapezoidal method (using M equispaced
%   intervals) if CHOICE=1, via the Gauss composite formula if CHOICE=2.
%   FUN accepts real scalar input x and returns a real scalar value.
%   FUN can also be an inline object.
H=(b-a)/M;
switch choice
case 1,
  x=linspace(a,b,M+1);
  fpm=feval(f,x,varargin{:});
  fpm(2:end-1)=2*fpm(2:end-1);
```

```
Itpc=0.5*H*sum(fpm);
case 2,
 z=linspace(a,b,M+1);
 x=z(1:end-1)+H/2*(1-1/sqrt(3));
 x=[x, x+H/sqrt(3)];
 fpm=feval(f,x,varargin{:});
 Itpc=0.5*H*sum(fpm);
otherwise
 disp('Unknown formula');
end
```

Example 4.2 We want to compare the approximations of the integral $I(f) = \int_0^{2\pi} xe^{-x} \cos(2x)\ dx = -0.122122604618968$, obtained by using the composite midpoint, trapezoidal and Gauss formulae. In Figure 4.4 we plot on the logarithmic scale the errors that are obtained versus H. As noticed in Section 1.5, in this type of plot the greater the slope of the curve, the higher the order of convergence of the corresponding formula. As expected from the theoretical results, the midpoint and trapezoidal formulae are second-order accurate, whereas the Gauss formula is fourth-order accurate.

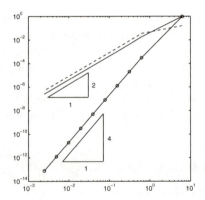

Fig. 4.4. Logarithmic representation of the errors versus H for Gauss (solid line with circles), midpoint (solid line) and trapezoidal (dashed line) composite quadrature formulae

 See the Exercises 4.9-4.11.

4.2.3 Simpson formula

The Simpson formula can be obtained by replacing the integral of f over each I_k by that of its interpolating polynomial of degree 2 at the nodes

x_{k-1}, $\bar{x}_k = (x_{k-1} + x_k)/2$ and x_k,

$$\Pi_2 f(x) = \frac{2(x - \bar{x}_k)(x - x_k)}{H^2} f(x_{k-1})$$
$$+ \frac{4(x_{k-1} - x)(x - x_k)}{H^2} f(\bar{x}_k) + \frac{2(x - \bar{x}_k)(x - x_{k-1})}{H^2} f(x_k).$$

The resulting formula is called the *Simpson composite quadrature formula*, and reads

$$I_s^c(f) = \frac{H}{6} \sum_{k=1}^{M} [f(x_{k-1}) + 4f(\bar{x}_k) + f(x_k)] \tag{4.21}$$

One can prove that it induces the error

$$I(f) - I_s^c(f) = -\frac{b - a}{180} \frac{H^4}{16} f^{(4)}(\xi), \tag{4.22}$$

where ξ is suitable point in (a, b), provided that $f \in C^4(a, b)$. It is therefore fourth-order accurate with respect to H. When (4.21) is applied to only one interval, say (a, b), we obtain the so-called *Simpson quadrature formula*

$$I_s(f) = \frac{b - a}{6} [f(a) + 4f((a + b)/2) + f(b)] \tag{4.23}$$

The error now is given by

$$I(f) - I_s(f) = -\frac{1}{16} \frac{(b - a)^5}{180} f^{(4)}(\xi), \tag{4.24}$$

for a suitable $\xi \in (a, b)$. Its degree of exactness is therefore equal to 3.

The composite Simpson rule is implemented in Program 5.

Program 5 - simpsonc: composite Simpson quadrature formula

```
function [Isic]=simpsonc(a,b,M,f,varargin)
%SIMPSONC Composite Simpson numerical integration.
%   ISIC = SIMPSONC(A,B,M,FUN) computes an approximation of the integral
%   of the function FUN via the Simpson method (using M equispaced intervals).
%   FUN accepts real scalar input x and returns a real scalar value.
%   FUN can also be an inline object.
```

```
H=(b-a)/M;
x=linspace(a,b,M+1);
fpm=feval(f,x,varargin{:});
fpm(2:end-1) = 2*fpm(2:end-1);
Isic=H*sum(fpm)/6;
x=linspace(a+H/2,b-H/2,M);
fpm=feval(f,x,varargin{:});
Isic = Isic+2*H*sum(fpm)/3;
```

Remark 4.1 (Interpolatory quadratures) All quadrature formulae introduced thus far are remarkable instances of a more general quadrature formula of the form:

$$I_{appr}(f) = \sum_{j=1}^{J} \alpha_j f(y_j) \qquad (4.25)$$

The real numbers α_j are the *quadrature weights*, while the points y_j are the *quadrature nodes*. In general, one requires that (4.25) integrates exactly at least a constant function: this property is ensured if $\sum_{j=1}^{J} \alpha_j = (b-a)$.

By properly chosing the weights and the nodes, the order of accuracy of formula (4.25) can be made arbitrarily large. For instance, using the MAT-
quadl   LAB instruction quadl(fun,a,b) (or the obsolete one quad8), it is possible
quad8   to compute an integral with a high order Gauss-Lobatto quadrature formula. The function fun can be an inline object. For instance, to integrate $f(x) = 1/x$ over $[1, 2]$, we must first define the following function fun

```
fun=inline('1./x','x');
```

then call quadl(fun,1,2). Note that in the definition of the function f we have used an element by element operation (indeed MATLAB will evaluate this expression component by component on the vector of quadrature nodes).

The specification of the number of subintervals is not requested as it is automatically computed in order to ensure that the quadrature error is below the default tolerance of 10^{-3}. A different tolerance can be provided by the user through the extended command quadl(fun,a,b,tol). In Section 4.3 we will introduce a method to estimate the quadrature error and, consequently, to change H adaptively.

Let us summarize

1. A quadrature formula is a formula to approximate the integral of continuous functions;

2. it is generally expressed as a linear combination of the values of the function at specific points (called *nodes*) with coefficients which are called *weights*;

3. the *degree of exactness* of a quadrature formula is the highest degree of the polynomials which are integrated exactly by the formula. The degree of exactness is 1 for midpoint and trapezoidal formulae, 3 for Gauss and Simpson formulae;

4. the *order of accuracy* of a composite quadrature formula is its order with respect to the size H of the subintervals. The order of accuracy is 2 for composite midpoint and trapezoidal formulae, 4 for composite Gauss and Simpson formulae.

See the Exercises 4.12-4.18.

4.3 Simpson adaptive formula

The integration step-length H of a quadrature composite formula can be chosen in order to ensure that the quadrature error is less than a prescribed tolerance $\varepsilon > 0$. For instance, when using the Simpson composite formula, this goal can be achieved by virtue of (4.22), if

$$\frac{b-a}{180}\frac{H^4}{16}\max_{x\in[a,b]}|f^{(4)}(x)| < \varepsilon, \tag{4.26}$$

where $f^{(4)}$ denotes the fourth-order derivative of f. Unfortunately, when the absolute value of $f^{(4)}$ is large only in a small part of the integration interval, the maximum H for which (4.26) holds true can be too small. The goal of the adaptive Simpson quadrature formula is to yield an approximation of $I(f)$ within a fixed tolerance ε by a *non uniform* distribution of the integration step-sizes in the interval $[a,b]$. In such a way we retain the same accuracy of the composite Simpson rule, but with a lower number of quadrature nodes and, consequently, a reduced number of evaluations of f.

To this end, we must find an error estimator and an automatic procedure to modify the integration step-length H, according to the achievement of the prescribed tolerance. We start by analyzing this procedure, which is independent of the specific quadrature formula that one wants to apply.

In the first step of the adaptive procedure, we compute an approximation $I_s(f)$ of $I(f) = \int_a^b f(x)\,dx$. We set $H = b-a$ and we try to estimate the quadrature error. If the error is less than the prescribed tolerance, the adaptive procedure is stopped; otherwise the step-size H is halved until the integral $\int_a^{a+H} f(x)\ dx$ is computed with the prescribed accuracy. When the test is passed, we consider the interval $(a + H, b)$ and we repeat the previous procedure, choosing as the first step-size the length $b - (a + H)$ of that interval.

We use the following notations:

1. A: the *active* integration interval, i.e., the interval where the integral is being computed;

2. S: the integration interval already examined, for which the error is less than the prescribed tolerance;

3. N: the integration interval yet to be examined.

At the beginning of the integration process we have $N = [a, b]$, $A = N$ and $S = \emptyset$, while the situation at the generic step of the algorithm is depicted in Figure 4.5. Let $J_S(f)$ indicate the computed approximation of $\int_a^\alpha f(x)dx$, with $J_S(f) = 0$ at the beginning of the process; if the algorithm successfully terminates, $J_S(f)$ yields the desired approximation of $I(f)$. We also denote by $J_{(\alpha,\beta)}(f)$ the approximate integral of f over the active interval $[\alpha, \beta]$. This interval is drawn in gray in Figure 4.5. The generic step of the adaptive integration method is organized as follows:

1. If the estimation of the error ensures that the prescribed tolerance is satisfied, then:

(i) $J_S(f)$ is increased by $J_{(\alpha,\beta)}(f)$, that is $J_S(f) \leftarrow J_S(f) + J_{(\alpha,\beta)}(f)$;

(ii) we let $S \leftarrow S \cup A$, $A = N$ (corresponding to the path (I) in Figure 4.5) and $\alpha \leftarrow \beta$ and $\beta \leftarrow b$;

2. If the estimation of the error fails the prescribed tolerance, then:

(j) A is halved, and the new active interval is set to $A = [\alpha, \alpha']$ with $\alpha' = (\alpha + \beta)/2$ (corresponding to the path (II) in Figure 4.5);

(jj) we let $N \leftarrow N \cup [\alpha', \beta]$, $\beta \leftarrow \alpha'$;

(jjj) a new error estimate is provided.

Of course, in order to prevent the algorithm from generating too small step-sizes, it is convenient to monitor the width of A and warn the user, in case of an excessive reduction of the step-length, about the presence of a possible singularity in the integrand function.

The problem now is to find a suitable estimator of the error. To this end, it is convenient to restrict our attention to a generic subinterval

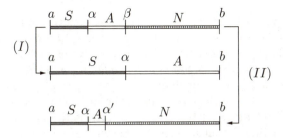

Fig. 4.5. Distribution of the integration intervals at the generic step of the adaptive algorithm and updating of the integration grid

$[\alpha, \beta]$ in which we compute $I_s(f)$: of course, if on this interval the error is less than $\varepsilon(\beta - \alpha)/(b - a)$, then the error on the interval $[a, b]$ will be less than the prescribed tolerance ε. Since from (4.24) we get

$$\int_\alpha^\beta f(x)\,dx - I_s(f) = -\frac{(\beta - \alpha)^5}{2880} f^{(4)}(\xi) = E_s(f; \alpha, \beta),$$

to ensure the achievement of the tolerance, it will be sufficient to verify that $E_s(f; \alpha, \beta) < \varepsilon(\beta - \alpha)/(b - a)$. In practical computation, this procedure is not feasible since the point $\xi \in [\alpha, \beta]$ is unknown.

To estimate the error without using explicitly the value $f^{(4)}(\xi)$, we employ again the composite Simpson formula to compute $\int_\alpha^\beta f(x)\,dx$, but with a step-length $(\beta - \alpha)/2$. From (4.22) with $a = \alpha$ and $b = \beta$, we deduce that

$$\int_\alpha^\beta f(x)\,dx - I_s^c(f) = -\frac{(\beta - \alpha)^5}{46080} f^{(4)}(\eta), \tag{4.27}$$

where η is a suitable point different from ξ. Subtracting the last two equations, we get

$$\Delta I = I_s^c(f) - I_s(f) = -\frac{(\beta - \alpha)^5}{2880} f^{(4)}(\xi) + \frac{(\beta - \alpha)^5}{46080} f^{(4)}(\eta). \tag{4.28}$$

Let us now make the assumption that $f^{(4)}(x)$ is approximately a constant on the interval $[\alpha, \beta]$. In this case $f^{(4)}(\xi) \simeq f^{(4)}(\eta)$. We can compute $f^{(4)}(\eta)$ from (4.28) and, putting this value in the equation (4.27), we obtain the following estimation of the error:

$$\int_\alpha^\beta f(x)\,dx - I_s^c(f) \simeq \frac{1}{15}\Delta I.$$

The step-length $(\beta-\alpha)/2$ (that is the step-length employed to compute $I_s^c(f)$) will be accepted if $|\Delta I|/15 < \varepsilon(\beta-\alpha)/[2(b-a)]$. The quadrature formula that uses this criterion in the adaptive procedure described previously, is called *adaptive Simpson formula*. It is implemented in Program 6. Among the input parameters, f is the string in which the function f is defined, a and b are the endpoints of the integration interval, tol is the prescribed tolerance on the error and hmin is the minimum admissible value of the integration step-length (in order to ensure that the adaption procedure always terminates).

Program 6 - simpadpt: adaptive Simpson formula

```
function [JSf,nodes]=simpadpt(f,a,b,tol,hmin)
%SIMPADPT   Numerically evaluate integral, adaptive Simpson quadrature.
%   JSF = SIMPADPT(FUN,A,B,TOL,HMIN) tries to approximate the
%   integral of function FUN from A to B to within an error of TOL using
%   recursive adaptive Simpson quadrature.  The inline function Y = FUN(V)
%   should accept a vector argument V and return a vector result Y, the
%   integrand evaluated at each element of X.
%
%   [JSF,NODES] = SIMPADPT(...) returns the distribution of nodes.
A=[a,b]; N=[]; S=[]; JSf = 0; ba = b - a; nodes=[];
while ~isempty(A),
  [deltal,ISc]=caldeltai(A,f);
  if abs(deltal) <= 15*tol*(A(2)-A(1))/ba;
     JSf = JSf + ISc;
     S = union(S,A);
     nodes = [nodes, A(1) (A(1)+A(2))*0.5 A(2)];
     S = [S(1), S(end)]; A = N; N = [];
  elseif A(2)-A(1) < hmin
     JSf=JSf+ISc;
     S = union(S,A);
     S = [S(1), S(end)]; A=N; N=[];
     warning('Too small step-length');
  else
     Am = (A(1)+A(2))*0.5;
     A = [A(1) Am];
     N = [Am, b];
  end
end
nodes=unique(nodes);
return

function [deltal,ISc]=caldeltai(A,f)
L=A(2)-A(1);
```

```
t=[0; 0.25; 0.5; 0.5; 0.75; 1];
x=L*t+A(1);
L=L/6;
w=[1; 4; 1];
fx=feval(f,x);
IS=L*sum(fx([1 3 6]).*w);
ISc=0.5*L*sum(fx.*[w;w]);
deltaI=IS-ISc;
return
```

Example 4.3 Let us compute the integral $I(f) = \int_{-1}^{1} e^{-10(x-1)^2} \, dx$ by the adaptive Simpson formula. Running Program 6 with $\texttt{tol} = 10^{-4}$ and $\texttt{hmin} = 10^{-3}$ provides 0.28024765884708, instead of the exact value 0.28024956081990. The error is less than the prescribed tolerance $\texttt{tol}=10^{-5}$. To obtain this result it was sufficient to use only 10 nonuniform subintervals. Note that the corresponding composite formula with uniform step-size would have required 22 subintervals to ensure the same accuracy.

4.4 What we haven't told you

The midpoint, trapezoidal and Simpson formulae are particular cases of a larger family of quadrature rules known as *Newton-Cotes formulae*. For an introduction, see [QSS00], Chapter 10. Analogously, the Gauss formula (4.20) is a special case of the important family of Gaussian quadrature formulae. These are *optimal* in the sense that they maximize the degree of exactness for a given number of quadrature nodes. In MATLAB 6 the function quadl implements one of these formulae. For an introduction to the Gaussian formulae, see [QSS00], Chapter 10 or [RR85]. Further developments on numerical integration can be found, *e.g.*, in [DR75] and [PdDKÜK83].

Numerical integration can also be used to compute integrals on unbounded intervals. For instance, to approximate $\int_{0}^{\infty} f(x) \, dx$, a first possibility is to find a point α such that the value of $\int_{\alpha}^{\infty} f(x) dx$ can be neglected with respect to that of $\int_{0}^{\alpha} f(x) dx$. Then we compute by a quadrature formula this latter integral on a bounded interval. A second possibility is to resort to Gaussian quadrature formulae for unbounded intervals (see [QSS00], Chapter 10).

Finally, numerical integration can also be used to compute multidimensional integrals. In particular, we mention the MATLAB instruction dblquad('f',xmin,xmax,ymin,ymax) by which it is possible to compute the integral of a function contained in the MATLAB file f.m over the rectangular domain [xmin,xmax] × [ymin,ymax]. Note that the

dblquad

function f must have at least two input parameters corresponding to the variables x and y with respect to which the integral is computed.

4.5 Exercises

Exercise 4.1 Verify that, if $f \in C^3$ in a neighborhood I_0 of x_0 (respectively, I_n of x_n) the error of the formula (4.10) is equal to $-\frac{1}{3}f'''(\xi_0)h^2$ (respectively, $-\frac{1}{3}f'''(\xi_n)h^2$), where ξ_0 and ξ_n are two suitable points belonging to I_0 and I_n, respectively.

Exercise 4.2 Verify that if $f \in C^3$ in a neighborhood of $\bar{x}$ the error of the formula (4.8) is equal to (4.9).

Exercise 4.3 Compute the order of accuracy with respect to h of the following formulae for the numerical approximation of $f'(x_i)$:

a. $$\frac{-11f(x_i) + 18f(x_{i+1}) - 9f(x_{i+2}) + 2f(x_{i+3})}{6h},$$

b. $$\frac{f(x_{i-2}) - 6f(x_{i-1}) + 3f(x_i) + 2f(x_{i+1})}{6h},$$

c. $$\frac{f(x_{i-2}) - 8f(x_i) + 8f(x_{i+1}) - f(x_{i+2})}{12h}.$$

Exercise 4.4 The following values represent the time evolution of the number $n(t)$ of individuals of a given population whose birth rate is constant ($b = 2$) and mortality rate is $d(t) = 0.01n(t)$:

t (months)	0	0.5	1	1.5	2	2.5	3
n	100	147	178	192	197	199	200

Use these data to approximate as accurately as possible the rate of variation of this population. Then compare the obtained results with the exact rate $n'(t) = 2n(t) - 0.01n^2(t)$.

Exercise 4.5 Find the minimum number M of subintervals to approximate with an absolute error less than 10^{-4} the integrals of the following functions:

$$f_1(x) = \frac{1}{1 + (x - \pi)^2} \quad \text{over } (0,5), \quad f_2(x) = e^x \cos(x) \quad \text{over } (0, \pi),$$

$$f_3(x) = \frac{1}{\sqrt{x(1 - x)}} \quad \text{over } (0, 1),$$

using the midpoint composite formula. Verify the results obtained using the Program 3.

Exercise 4.6 Prove (4.13) starting from (4.14).

Exercise 4.7 Why does the midpoint formula lose one order of convergence when used in its composite mode?

Exercise 4.8 Verify that, if f is a polynomial of degree less than or equal 1, then $I_{mp}(f) = I(f)$ i.e. the midpoint formula has degree of exactness equal to 1.

Exercise 4.9 For the function f_1 of Exercise 4.5, compute (numerically) the values of M which ensure that the quadrature error is less than 10^{-4} when the integral is approximated by the composite trapezoidal and Gauss quadrature formulae.

Exercise 4.10 Let I_1 and I_2 be two values obtained by the composite trapezoidal formula applied with two different step-lengths, H_1 and H_2, for the approximation of $I(f) = \int_a^b f(x)dx$. Verify that, if $f^{(2)}$ has a mild variation on (a, b), the value

$$I_R = I_1 + (I_1 - I_2)/(H_2^2/H_1^2 - 1) \tag{4.29}$$

is an approximation of $I(f)$ better than I_1 and I_2. This strategy is called the *Richardson extrapolation method*.

Exercise 4.11 Verify that, among all formulae of the form $I_{appx}(f) = \alpha f(\bar{x}) + \beta f(\bar{z})$ where $\bar{x}, \bar{z} \in [a, b]$ are two unknown nodes and α and β two undetermined weights, formula (4.20) features the maximum degree of exactness.

Exercise 4.12 For the first two functions of Exercise 4.5, compute the minimum number of intervals such that the quadrature error of the composite Simpson quadrature formula is less than 10^{-4}.

Exercise 4.13 Compute $\int_0^2 e^{-x^2/2} \, dx$ by the Simpson formula (4.23) and Gauss formula (4.20), then compare the obtained results.

Exercise 4.14 To compute the integrals $I_k = \int_0^1 x^k e^{x-1} dx$ for $k = 1, 2, \ldots$, one can use the following recursive formula: $I_k = 1 - kI_{k-1}$, with $I_1 = 1/e$. Compute I_{20} by the composite Simpson formula in order to ensure that the quadrature error is less than 10^{-3}. Compare the Simpson approximation with the result obtained by the above recursive formula.

Exercise 4.15 Apply the Richardson extrapolation formula (4.29) for the approximation of the integral $I(f) = \int_0^2 e^{-x^2/2} dx$, with $H_1 = 1$ and $H_2 = 0.5$ using first Simpson formula (4.23), then the Gauss formula (4.20). Verify that I_R is more accurate than I_1 and I_2.

Exercise 4.16 Compute by the composite Simpson formula the function $j(r)$ defined in (4.2) for $r = k/10$ m with $k = 1, \ldots, 10$, with $\rho(\xi) = e^\xi$ and $\sigma = 0.36$ W/(mK). Ensure that the quadrature error is less than 10^{-10}. (Recall that: m=meters, W=watts, K=degrees Kelvin.)

Exercise 4.17 By using the composite Simpson and Gauss formulae compute the function $E(T)$, defined in (4.1), for T equal to 213 K, up to at least 10 exact significant digits.

Exercise 4.18 Develop a strategy to compute $I(f) = \int_1^0 |x^2 - 0.25|\, dx$ by the composite Simpson formula such that the quadrature error is less than 10^{-2}.

5. Linear systems

In applied sciences, one is quite often led to face a linear system of the form

$$\mathbf{A}\mathbf{x} = \mathbf{b} \qquad (5.1)$$

for the solution of complex problems, where A is a square matrix of dimension $n \times n$ whose elements a_{ij} are either real or complex, while $\mathbf{x}$ and $\mathbf{b}$ are column vectors of dimension n with $\mathbf{x}$ representing the unknown solution and $\mathbf{b}$ is a given vector. Component-wise, (5.1) can be written as

$$a_{11}x_1 + a_{12}x_2 + \ldots + a_{1n}x_n = b_1,$$

$$a_{21}x_1 + a_{22}x_2 + \ldots + a_{2n}x_n = b_2,$$

$$\vdots \qquad\qquad \vdots \quad \vdots$$

$$a_{n1}x_1 + a_{n2}x_2 + \ldots + a_{nn}x_n = b_n.$$

We present three different problems that give rise to linear systems.

Problem 5.1 (Hydraulics) Let us consider the hydraulic network made of the 10 pipelines in Figure 5.1, which is fed by a reservoir of water at constant pressure $p_r = 10$ bar. In this problem, pressure values refer to the difference between the real pressure and the atmospheric one. For the j-th pipeline, the following relationship holds between the flow-rate Q_j (in m^3/s) and the pressure gap Δp_j at pipe-ends:

$$Q_j = kL\Delta p_j, \qquad (5.2)$$

where k is the hydraulic resistance (in m^2 /(bar s)) and L is the length (in m) of the pipeline. We assume that water flows from the outlets (indicated by a black dot) at atmospheric pressure, which is set to 0 bar for coherence with the previous convention.

A typical problem consists of determining the pressure values at every internal node 1, 2, 3, 4. With this aim, for each $j = 1, 2, 3, 4$ we can supplement the relationship (5.2) with the statement that the algebraic sum of the flow-rates of the pipelines which meet at j-th node must be null (a negative value would indicate the presence of a seepage).

Denoting by $\mathbf{p} = (p_1, p_2, p_3, p_4)^T$ the pressure vector at the internal nodes, we get a 4×4 system of the form $A\mathbf{p} = \mathbf{b}$.

Let us summarize in the following table the relevant characteristics of the different pipelines.

pipeline	k	L	pipeline	k	L	pipeline	k	L
1	0.01	20	2	0.005	10	3	0.005	14
4	0.005	10	5	0.005	10	6	0.002	8
7	0.002	8	8	0.002	8	9	0.005	10
10	0.002	8						

Correspondingly, A and **b** take the following values (only the first 4 significant digits are provided):

$$
A = \begin{bmatrix} -0.370 & 0.050 & 0.050 & 0.070 \\ 0.050 & -0.116 & 0 & 0.050 \\ 0.050 & 0 & -0.116 & 0.050 \\ 0.070 & 0.050 & 0.050 & -0.202 \end{bmatrix}, \quad \mathbf{b} = \begin{bmatrix} -2 \\ 0 \\ 0 \\ 0 \end{bmatrix}.
$$

The solution of this system is postponed to Example 5.4.   •

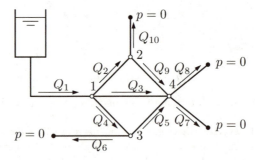

Fig. 5.1. The pipeline network of Problem 5.1

Problem 5.2 (Spectrometry) Let us consider a gas mixture of n non reactive unknown components. By the mass spectrometer the compound is bombarded by low-energy electrons: the resulting mixture of ions is analyzed by a galvanometer which shows peaks corresponding to specific ratios mass/charge. We consider only the n most relevant peaks. One may conjecture that the height h_i of the i-th peak is a linear combination of $\{p_j, j = 1, \ldots, n\}$, p_j

being the partial pressure of the j-th component (that is the pressure exerted by a single gas when it is part of a mixture), yielding

$$\sum_{j=1}^{n} s_{ij} p_j = h_i, \quad i = 1, \ldots, n,$$

where the s_{ij} are the so-called sensitivity coefficients. The determination of the partial pressures demands therefore the solution of a linear system.   ●

Problem 5.3 (Economy: input-output analysis) We want to determine the situation of equilibrium between demand and offer of certain goods. In particular, let us consider n factories that produce n different products. They must face the internal demand of goods necessary to the factories for their own production, as well as the external demand from the consumers. Let x_i, $i = 1, \ldots, n$, denote the number of units of the i-th product of the i-th factory and let b_i, $i = 1, \ldots, n$, denote the number of units of the i-th product absorbed by the market. Finally, let c_{ij} be the fraction of x_i necessary to the j-th factory for the production of the j-th product (see Figure 5.2). According to the Leontief model, we assume that the transformation functions that relate the various problems are linear. Then the equilibrium is reached when the vector $\mathbf{x}$ of the total production equals the total demand, that is, $\mathbf{x} = C\mathbf{x} + \mathbf{b}$, where $C = (c_{ij})$ and $\mathbf{b} = (b_i)$. Thus, in this model the total production satisfies the linear system

$$A\mathbf{x} = \mathbf{b}, \quad \text{where } A = I - C. \tag{5.3}$$

For its solution, see Exercise 5.17.   ●

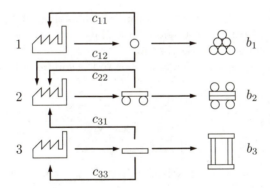

Fig. 5.2. The interaction scheme of 3 factories with the market

The solution of the system (5.1) exists iff A is non singular. In principle, the solution might be computed by the so-called *Cramer rule*:

$$x_i = \frac{\det(A_i)}{\det(A)}, \quad i = 1, \ldots, n,$$

where A_i is the matrix obtained from A by replacing the i-th column by **b** and $\det(A)$ denotes the determinant of A. If the $n+1$ determinants are computed by the Laplace expansion (see Exercise 5.1), a total number of $2(n + 1)!$ *ops* is required. As usual, by operation we mean a sum, a subtraction, a product or a division. For instance, a computer capable of carrying out 10^9 *flops* (i.e. 1 giga *ops* per second), would require about 12 hours to solve a system of dimension $n = 15$, 3240 years if $n = 20$ and 10^{143} years if $n = 100$. The computational cost can be drastically reduced to the order of about $n^{3.8}$ *ops* if the $n + 1$ determinants are computed by the algorithm quoted in Example 1.3. Yet, this cost is still too high for the large values of n which often arise in practical applications.

We need therefore to look for alternative approaches. Let us note that in general a system cannot be solved by less than n^2 *ops*. Indeed, if the equations are fully coupled, we should expect that every one of the n^2 matrix coefficients would be involved in an operation at least once.

5.1   The LU factorization method

Let A be a square matrix of order n. Assume that there exist two suitable matrices L and U, lower triangular and upper triangular, respectively, such that

$$A = LU. \qquad (5.4)$$

We call (5.4) an LU-*factorization* (or decomposition) of A. Should A be non-singular, so are both L and U, and thus their diagonal elements are non-null (as observed in Section 1.3).

In such a case, solving $A\mathbf{x} = \mathbf{b}$ leads to the solution of two triangular systems

$$\boxed{L\mathbf{y} = \mathbf{b}, \quad U\mathbf{x} = \mathbf{y}.} \qquad (5.5)$$

Both systems are easy to solve. Indeed, L being lower triangular, the first row of the system $L\mathbf{y} = \mathbf{b}$ takes the form:

$$l_{11}y_1 = b_1,$$

which provides the value of y_1 since $l_{11} \neq 0$. By substituting this value of y_1 in the subsequent $n - 1$ equations we obtain a new system whose unknowns are $y_2, \ldots, y_n$, on which we can proceed in a similar manner.

Proceeding forward, equation by equation, we can compute all unknowns with the following *forward substitutions* algorithm:

$$
\begin{aligned}
y_1 &= \frac{1}{l_{11}} b_1, \\
y_i &= \frac{1}{l_{ii}} \left(b_i - \sum_{j=1}^{i-1} l_{ij} y_j \right), \quad i = 2, \ldots, n
\end{aligned}
\tag{5.6}
$$

Let us quantify the number of operations required from (5.6). Since $i - 1$ sums, $i - 1$ products and 1 division are needed to compute the unknown y_i, the total number of operations required is

$$
\sum_{i=1}^{n} 1 + 2 \sum_{i=1}^{n} (i - 1) = 2 \sum_{i=1}^{n} i - n = n^2.
$$

The system $\mathbf{Ux} = \mathbf{y}$ can be solved by proceeding in a similar manner. This time, the first unknown to be computed is x_n then, by proceeding backward, we can compute the remaining unknowns x_i, for $i = n - 1$ to $i = 1$:

$$
\begin{aligned}
x_n &= \frac{1}{u_{nn}} y_n, \\
x_i &= \frac{1}{u_{ii}} \left(y_i - \sum_{j=i+1}^{n} u_{ij} x_j \right), \quad i = n - 1, \ldots, 1
\end{aligned}
\tag{5.7}
$$

This is called *backward substitutions* algorithm and requires n^2 *ops* too. At this stage we need an algorithm that allows an effective computation of the factors L and U of the matrix A. We illustrate a general procedure starting from a couple of examples.

Example 5.1 Let us write the relation (5.4) for a generic matrix $A \in \mathbb{R}^{2 \times 2}$

$$
\begin{bmatrix} l_{11} & 0 \\ l_{21} & l_{22} \end{bmatrix} \begin{bmatrix} u_{11} & u_{12} \\ 0 & u_{22} \end{bmatrix} = \begin{bmatrix} a_{11} & a_{12} \\ a_{21} & a_{22} \end{bmatrix}.
$$

The 6 unknown elements of L and U must satisfy the following (non-linear) equations:

$$
\begin{aligned}
&(e_1)\ l_{11} u_{11} = a_{11}, \quad (e_2)\ l_{11} u_{12} = a_{12}, \\
&(e_3)\ l_{21} u_{11} = a_{21}, \quad (e_4)\ l_{21} u_{12} + l_{22} u_{22} = a_{22}.
\end{aligned}
\tag{5.8}
$$

System (5.8) is *underdetermined* as it features less equations than unknowns. We can complete it by assigning *arbitrarily* the diagonal elements of L, for instance setting $l_{11} = 1$ and $l_{22} = 1$. Now system (5.8) can be solved by proceeding as follows: we determine the elements u_{11} and u_{12} of the first row of U using (e_1) and (e_2). If u_{11} is nonnull then from (e_3) we deduce l_{21} (that is the first column of L, since l_{11} is already available). Now we can obtain from (e_4) the only non zero element u_{22} of the second row of U.

Example 5.2 Let us repeat the same computations in the case of a 3×3 matrix. For the 12 unknown coefficients of L and U we have the following 9 equations:

$$(e_1)\ l_{11}u_{11} = a_{11}, \quad (e_2)\ l_{11}u_{12} = a_{12}, \quad\quad (e_3)\ l_{11}u_{13} = a_{13},$$
$$(e_4)\ l_{21}u_{11} = a_{21}, \quad (e_5)\ l_{21}u_{12} + l_{22}u_{22} = a_{22}, \quad (e_6)\ l_{21}u_{13} + l_{22}u_{23} = a_{23},$$
$$(e_7)\ l_{31}u_{11} = a_{31}, \quad (e_8)\ l_{31}u_{12} + l_{32}u_{22} = a_{32}, \quad (e_9)\ l_{31}u_{13} + l_{32}u_{23} + l_{33}u_{33}$$
$$= a_{33}.$$

Let us complete this system by setting $l_{ii} = 1$ for $i = 1, 2, 3$. Now, the coefficients of the first row of U can be obtained by using (e_1), (e_2) and (e_3). Next, using (e_4) and (e_7), we can determine the coefficients l_{21} and l_{31} of the first column of L. Using (e_5) and (e_6) we can now compute the coefficients u_{22} and u_{23} of the second row of U. Then, using (e_8), we obtain the coefficient l_{32} of the second column of L. Finally, the last row of U (which consists of the only element u_{33}) can be determined by solving (e_9).

On a matrix of arbitrary dimension n we can proceed as follows:

1. The elements of L and U satisfy the system of non-linear equations

$$\sum_{r=1}^{\min(i,j)} l_{ir}u_{rj} = a_{ij}, \quad i, j = 1, \dots, n; \tag{5.9}$$

2. System (5.9) is underdetermined; indeed there are n^2 equations and $n^2 + n$ unknowns, thus the factorization LU cannot be unique;

3. By forcing the n diagonal elements of L to be equal to 1, (5.9) turns into a determined system which can be solved by the following *Gauss algorithm*: set $A^{(1)} = A$ *i.e.* $a_{ij}^{(1)} = a_{ij}$ for $i, j = 1, \dots, n$; for $k = 1, \dots, n-1$ do

$$\boxed{\begin{aligned} &\text{for } i = k+1, \dots, n \\ &\quad l_{ik} = \frac{a_{ik}^{(k)}}{a_{kk}^{(k)}}, \\ &\quad \text{for } j = k+1, \dots, n \\ &\quad\quad a_{ij}^{(k+1)} = a_{ij}^{(k)} - l_{ik}a_{kj}^{(k)} \end{aligned}} \tag{5.10}$$

The elements $a_{kk}^{(k)}$ must all be different from zero and are called *pivot* elements. For every $k = 1, \ldots, n-1$ the matrix $A^{(k+1)} = (a_{ij}^{(k+1)})$ has $n - k$ rows and columns.

At the end of this procedure the elements of the upper triangular matrix U are given by $u_{ij} = a_{ij}^{(i)}$ for $i = 1, \ldots, n$ and $j = i, \ldots, n$, whereas those of L are given by the coefficients l_{ij} generated by this algorithm. In (5.10) there is no computation of the diagonal elements of L, as we already know that their value is equal to 1.

This factorization is called the *Gauss factorization*; the determination of the elements of the factors L and U requires about $2n^3/3$ *ops* (see Exercise 5.4).

Example 5.3 Consider the following Vandermonde matrix

$$A = (a_{ij}) \quad \text{with } a_{ij} = x_i^{(n-j)}, \ i, j = 1, \ldots, n, \tag{5.11}$$

where the x_i are n distinct abscissae. It can be constructed using the MAT-LAB command **vander**. In Figure 5.3 we report the number of floating point   vander operations required in order to compute the Gauss factorization of A. Several values of n are considered in abscissae, and the corresponding number of *ops* are indicated with circles. They have been obtained using the following commands:

```
>> ops = [ ]; for n = [10:10:100]
   A = vander(linspace(1,2,n));
   flops(0), A = lu_gauss(A);
   ops=[ops,flops];
end
```

The curve reported in the picture is a polynomial in n of the third order representing the least square approximation of the above data, and is generated by the following commands:

```
>> c3=polyfit([10:10:100],ops,3)
   0.6667   -0.0000    0.3333   -0.0000
```

The coefficient of the monomial n^3 is exactly $2/3$.

It is not necessary to store the matrices $\{A^{(k)}\}$; actually we can overlap the $(n-k) \times (n-k)$ elements of $A^{(k+1)}$ on the corresponding last $(n-k) \times (n-k)$ elements of the original matrix A. Moreover, since at the k-th step, the subdiagonal elements of the k-th column don't have any effect on the final U, they can be replaced by the entries of the k-th column of L, as done in Program 7. Then, at the k-th step of the process the

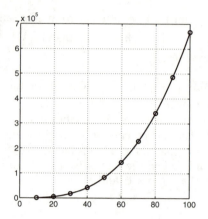

Fig. 5.3. The number of *ops* as a function of n, necessary to generate the Gauss factorization LU of the Vandermonde matrix. This function is a cubic polynomial obtained by approximating in the least square sense the values (represented by circles) corresponding to $n = 10, 20, \dots, 100$

elements stored at location of the original entries of A are

$$
\begin{bmatrix}
a_{11}^{(1)} & a_{12}^{(1)} & \cdots & & & & a_{1n}^{(1)} \\
l_{21} & a_{22}^{(2)} & & & & & a_{2n}^{(2)} \\
\vdots & & \ddots & \ddots & & & \vdots \\
l_{k1} & \cdots & l_{k,k-1} & & a_{kk}^{(k)} & \cdots & a_{kn}^{(k)} \\
\vdots & & \vdots & & \vdots & & \vdots \\
l_{n1} & \cdots & l_{n,k-1} & & a_{nk}^{(k)} & \cdots & a_{nn}^{(k)}
\end{bmatrix},
$$

where the boxed submatrix is $A^{(k)}$. The Gauss factorization is the basis of several MATLAB commands:

lu   [L,U]=lu(A)   whose mode of use will be discussed in Section 5.2;

inv   inv that allows the computation of the inverse of a matrix;

\   \ by which it is possible to solve a linear system with matrix A and right hand side b by simply writing A \ b.

Remark 5.1 (Computing a determinant) By means of the LU factorization one can compute the determinant of A with a computational cost of $\mathcal{O}(n^3)$ operations, noting that (see Sect.1.3)

$$
\det(A) = \det(L)\,\det(U) = \prod_{k=1}^{n} u_{kk}.
$$

As a matter of fact, this procedure is also at the basis of the MATLAB command **det**.

det

In Program 7 we implement the algorithm (5.10). The factor L is stored in the (strictly) lower triangular part of A and U in the upper triangular part of A (for the sake of storage saving). After the program execution, the two factors can be recovered by simply writing: L = eye(n) + tril(A,-1) and U = triu(A), where n is the size of A.

Program 7 - lu_gauss: Gauss factorization

```
function A=lu_gauss(A)
%LU_GAUSS    LU factorization without pivoting.
%   A = LU_GAUSS(A) stores an upper triangular matrix in the upper triangular
%   part of A and a lower triangular matrix in the strictly lower part of A
%   (the diagonal elements of L are 1).
[n,m]=size(A);
if n ~= m; error('A is not a square matrix'); else
 for k = 1:n-1
  for i = k+1:n
   A(i,k) = A(i,k)/A(k,k);
   if A(k,k) == 0, error('Null diagonal element'); end
   for j = k+1:n
    A(i,j) = A(i,j) - A(i,k)*A(k,j);
   end
  end
 end
end
```

Example 5.4 Let us compute the solution of the system encountered in the Problem 5.1 by using the LU factorization, then applying the backward and forward substitution algorithms. We need to compute the matrix A and the right hand side b and execute the following instructions:

```
>> A=lu_gauss(A);
>> y(1)=b(1); for i = 2:4; y = [y; b(i)-A(i,1:i-1)*y(1:i-1)]; end
>> x(4)=y(4)/A(4,4);
>> for i = 3:-1:1; x(i)= (y(i)-A(i,i+1:4)*x(i+1:4))/A(i,i);  end
```

The result is $\mathbf{p} = (8.1172, 5.9893, 5.9893, 5.7779)^T$.

Example 5.5 Suppose that we solve $A\mathbf{x} = \mathbf{b}$ with

$$A = \begin{bmatrix} 1 & 1-\varepsilon & 3 \\ 2 & 2 & 2 \\ 3 & 6 & 4 \end{bmatrix}, \quad \mathbf{b} = \begin{bmatrix} 5-\varepsilon \\ 6 \\ 13 \end{bmatrix}, \quad \varepsilon \in \mathbb{R}, \qquad (5.12)$$

whose solution is $\mathbf{x} = (1,1,1)^T$ (independently of the value of ε).

Let us set $\varepsilon = 1$. The Gauss factorization of A obtained by the Program 7 yields

$$L = \begin{bmatrix} 1 & 0 & 0 \\ 2 & 1 & 0 \\ 3 & 3 & 1 \end{bmatrix}, \quad U = \begin{bmatrix} 1 & 0 & 3 \\ 0 & 2 & -4 \\ 0 & 0 & 7 \end{bmatrix}.$$

If we set $\varepsilon = 0$, despite the fact that A is non singular, the Gauss factorization cannot be carried out since the algorithm (5.10) would involve divisions by 0.

The previous example shows that, unfortunately, the Gauss factorization A=LU does not necessarily exist for every non singular matrix A. In this respect, the following result can be proven:

Proposition 5.1 *For a given matrix* $A \in \mathbb{R}^{n \times n}$*, its Gauss factorization exists if and only if the principal submatrices* A_i *of* A *of order* $i = 1, \ldots, n - 1$ *(that is those obtained by restricting* A *to its first* i *rows and columns) are non singular. This factorization is unique if* A *is non singular.*

Going back to Example 5.5, we can notice that when $\varepsilon = 0$ the second principal submatrix A_2 of the matrix A is singular.

We can identify special classes of matrices for which the hypotheses of Proposition 5.1 are fulfilled. In particular, we mention:

1. Symmetric and positive definite matrices. A matrix $A \in \mathbb{R}^{n \times n}$ is *positive definite* if

$$\forall \mathbf{x} \in \mathbb{R}^n \text{ with } \mathbf{x} \neq \mathbf{0}, \qquad \mathbf{x}^T A \mathbf{x} > 0;$$

2. Diagonally dominant matrices. A matrix is *diagonally dominant by row* if

$$|a_{ii}| \geq \sum_{\substack{j=1 \\ j \neq i}}^{n} |a_{ij}|, \quad i = 1, \ldots, n,$$

by column if

$$|a_{ii}| \geq \sum_{\substack{j=1 \\ j \neq i}}^{n} |a_{ji}|, \quad i = 1, \ldots, n.$$

A special case occurs when in the previous inequalities we can replace $\geq$ by $>$. Then the matrix A is called *strictly* diagonally dominant (by row or by column, respectively).

If A is symmetric and positive definite, it is moreover possible to construct a special factorization:

$$A = HH^T, \tag{5.13}$$

where H is a lower triangular matrix with positive diagonal elements. This is the so-called *Cholesky factorization* and requires about $n^3/3$ operations (half of those required by the Gauss LU factorization). Further, let us note that, due to the symmetry, only the lower part of A is stored, and H can be stored in the same area.

The elements of H can be computed by the following algorithm: for $i = 1, \ldots, n$,

$$h_{ij} = \frac{1}{h_{jj}} \left(a_{ij} - \sum_{k=1}^{j-1} h_{ik} h_{jk} \right), \quad j = 1, \ldots, i-1,$$

$$h_{ii} = \sqrt{a_{ii} - \sum_{k=1}^{i-1} h_{ik}^2}.$$

Cholesky factorization is available in MATLAB by setting `H=chol(A)`. `chol`

See the Exercises 5.1-5.5.

5.2   The technique of pivoting

We are going to introduce a special technique that allows us to achieve the LU factorization for every non singular matrix, even if the hypotheses of Proposition 5.1 are not fulfilled.

Let us go back to the case described in Example 5.5 and take $\varepsilon = 0$. Setting $A^{(1)} = A$ after having carried out the first step ($k = 1$) of the procedure, the new entries of A are

$$\begin{bmatrix} \begin{array}{c|cc} 1 & 1 & 3 \\ \hline 2 & \mathbf{0} & -4 \\ 3 & 3 & -5 \end{array} \end{bmatrix}. \tag{5.14}$$

Since the *pivot* a_{22} is equal to zero, this procedure cannot be continued further. On the other hand, should we interchange the second and third rows beforehand, we would obtain the matrix

$$\begin{bmatrix} \begin{array}{c|cc} 1 & 1 & 3 \\ \hline 3 & \mathbf{3} & -5 \\ 2 & 0 & -4 \end{array} \end{bmatrix}$$

and thus the factorization could be accomplished without involving a division by 0.

We can state that *permutation* in a suitable manner of the rows of the original matrix A would make the entire factorization procedure feasible even if the hypotheses of Proposition 5.1 are not verified. Unfortunately, we cannot know *a priori* which rows should be permuted. However, this decision can be made at every step k at which a null diagonal element $a_{kk}^{(k)}$ is generated.

Let us return to the matrix $A^{(2)}$ in (5.14): since $a_{22}^{(2)}$ is null, let us interchange the third and second row of $A^{(2)}$ and check whether the new $a_{22}^{(2)}$ that is generated is still null. By executing the second step of the factorization procedure we find the same matrix that we would have generated by an *a priori* permutation of the same two rows of A.

We can therefore perform a row permutation as soon as this becomes necessary, without carrying out any *a priori* transformation on A. Since a row permutation entails changing the *pivot* element, this technique is given the name of *pivoting by row*. The factorization that is generated in this way returns the original matrix up to a row permutation. Precisely we obtain

$$PA = LU.$$

P is a suitable *permutation matrix*. At the beginning it is set equal to the identity matrix, then if in the course of the procedure the rows r and s of A are permuted, the same permutation must be performed on the homologous rows of P. Correspondingly, we should now solve the following triangular systems

$$Ly = Pb, \quad Ux = y. \tag{5.15}$$

From the second equation of (5.10) we see that not only null pivot elements $a_{kk}^{(k)}$ are troublesome, but also those which are very small. Indeed, should $a_{kk}^{(k)}$ be near zero, possible roundoff errors affecting the coefficients $a_{kj}^{(k)}$ will be severely amplified.

Example 5.6 Consider the nonsingular matrix

$$A = \begin{bmatrix} 1 & 1 + 0.5 \cdot 10^{-15} & 3 \\ 2 & 2 & 20 \\ 3 & 6 & 4 \end{bmatrix}.$$

During the factorization procedure by Program 7 no null pivot elements are obtained. Yet, the factors L and U turn out to be quite inaccurate, as one can realize by computing the residual matrix A − LU (which should be the null matrix if all operations were carried out in exact arithmetic):

$$A - LU = \begin{bmatrix} 0 & 0 & 0 \\ 0 & 0 & 0 \\ 0 & 0 & 6 \end{bmatrix}.$$

It is therefore recommended to carry out the *pivoting* at every step of the factorization procedure, by searching among all virtual *pivot* elements $a_{ik}^{(k)}$ with $i = k, \ldots, n$, the one with maximum modulus. The algorithm (5.10) with *pivoting* by row carried out at every step takes the following form: for $k = 1, \ldots, n$, do

> for $i = k + 1, \ldots, n$
> find $\bar{r}$ such that $|a_{\bar{r}k}^{(k)}| = \max_{r=k,\ldots,n} |a_{rk}^{(k)}|$,
> exchange row k with row $\bar{r}$,
> $$l_{ik} = \frac{a_{ik}^{(k)}}{a_{kk}^{(k)}},$$
> for $j = k + 1, \ldots, n$
> $$a_{ij}^{(k+1)} = a_{ij}^{(k)} - l_{ik} a_{kj}^{(k)}$$

(5.16)

The MATLAB program `lu` that we have mentioned previously computes the Gauss factorization with *pivoting* by row. Its complete syntax is indeed `[L,U,P]=lu(A)`, P being the permutation matrix. When called in the shorthand mode `[L,U]=lu(A)`, the matrix L is equal to `P*M`, where M is lower triangular and P is the permutation matrix generated by the *pivoting* by row. The program `lu` activates automatically the *pivoting* by row when a null (or very small) *pivot* element is computed.

See the Exercises 5.6-5.8.

5.3 How accurate is the LU factorization?

We have already noticed in Example 5.6 that, due to roundoff errors, the product LU does not reproduce A exactly. Even though the *pivoting* strategy damps these errors, yet the result could sometimes be rather unsatisfactory.

Example 5.7 Consider the linear system $A_n x_n = b_n$ where $A_n \in \mathbb{R}^{n \times n}$ is the so-called *Hilbert matrix* whose elements are

$$a_{ij} = 1/(i + j - 1), \qquad i, j = 1, \ldots, n,$$

while $\mathbf{b}_n$ is chosen in such a way that the exact solution is $\mathbf{x}_n = (1, 1, \ldots, 1)^T$. The matrix A_n is clearly symmetric and one can prove that it is also positive definite.

For different values of n we use the MATLAB function lu to get the Gauss factorization of A_n with *pivoting* by row. Then we solve the associated linear systems (5.15) and denote by $\widehat{\mathbf{x}}_n$ the computed solution. In Figure 5.4 we report (in logarithmic scale) the relative errors

$$E_n = \|\mathbf{x}_n - \widehat{\mathbf{x}}_n\| / \|\mathbf{x}_n\|, \tag{5.17}$$

having denoted by $\| \cdot \|$ the Euclidean norm introduced in the Section 1.3.1. We have $E_n \geq 10$ if $n \geq 13$ (that is a relative error on the solution higher than 1000%!), whereas $R_n = L_n U_n - P_n A_n$ is the null matrix (up to machine accuracy) for whatever value of n.

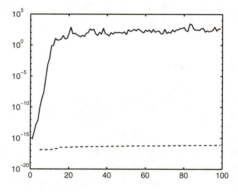

Fig. 5.4. Behavior versus n of E_n (solid line) and of $\max_{i,j=1,\ldots,n} |r_{ij}|$ (dashed line) in logarithmic scale, for the Hilbert system of Example 5.7. The (r_{ij}) are the coefficients of the matrix R

On the ground of the previous remark, we could speculate by saying that, when a linear system $A\mathbf{x} = \mathbf{b}$ is solved numerically, one is indeed looking for the *exact* solution $\widehat{\mathbf{x}}$ of a *perturbed* system

$$(A + \delta A)\widehat{\mathbf{x}} = \mathbf{b} + \boldsymbol{\delta b}, \tag{5.18}$$

where δA and $\boldsymbol{\delta b}$ are respectively a matrix and a vector which depend on the specific numerical method which is being used. We start by considering the case where $\delta A = 0$ and $\boldsymbol{\delta b} \neq \mathbf{0}$ which is simpler than the most general case. Moreover, for simplicity we will also assume that A is symmetric and positive definite.

By comparing (5.1) and (5.18) we find $\mathbf{x} - \widehat{\mathbf{x}} = -A^{-1}\boldsymbol{\delta b}$, and thus

$$\|\mathbf{x} - \widehat{\mathbf{x}}\| = \|A^{-1}\boldsymbol{\delta b}\|. \tag{5.19}$$

In order to find an upper bound for the right hand side of (5.19), we proceed as follows. Since A is symmetric and positive definite, the set of its eigenvectors $\{\mathbf{v}_i\}_{i=1}^n$ furnishes an orthonormal basis of $\mathbb{R}^n$ (see [QSS00], chapter 5). This means that

$$A\mathbf{v}_i = \lambda_i \mathbf{v}_i, \quad i = 1, \ldots, n,$$

$$\mathbf{v}_i^T \mathbf{v}_j = \delta_{ij}, \quad i, j = 1, \ldots, n,$$

where λ_i is the eigenvalue of A associated with $\mathbf{v}_i$ and δ_{ij} is the Kronecker symbol. Consequently, a generic vector $\mathbf{w} \in \mathbb{R}^n$ can be written as

$$\mathbf{w} = \sum_{i=1}^n w_i \mathbf{v}_i,$$

for a suitable (and unique) set of coefficients $w_i \in \mathbb{R}$. We have

$$
\begin{aligned}
\|A\mathbf{w}\|^2 &= (A\mathbf{w})^T(A\mathbf{w}) \\
&= [w_1(A\mathbf{v}_1)^T + \ldots + w_n(A\mathbf{v}_n)^T][w_1 A\mathbf{v}_1 + \ldots + w_n A\mathbf{v}_n] \\
&= (\lambda_1 w_1 \mathbf{v}_1^T + \ldots + \lambda_n w_n \mathbf{v}_n^T)(\lambda_1 w_1 \mathbf{v}_1 + \ldots + \lambda_n w_n \mathbf{v}_n) \\
&= \sum_{i=1}^n \lambda_i^2 w_i^2.
\end{aligned}
$$

Denote by λ_{max} the largest eigenvalue of A. Since $\|\mathbf{w}\|^2 = \sum_{i=1}^n w_i^2$, we conclude that

$$\|A\mathbf{w}\| \leq \lambda_{max}\|\mathbf{w}\| \quad \forall \mathbf{w} \in \mathbb{R}^n. \tag{5.20}$$

In a similar manner, we obtain

$$\|A^{-1}\mathbf{w}\| \leq \frac{1}{\lambda_{min}}\|\mathbf{w}\|,$$

upon recalling that the eigenvalues of A^{-1} are the reciprocals of those of A. This inequality enables us to draw from (5.19) that

$$\frac{\|\mathbf{x} - \widehat{\mathbf{x}}\|}{\|\mathbf{x}\|} \leq \frac{1}{\lambda_{min}} \frac{\|\boldsymbol{\delta}\mathbf{b}\|}{\|\mathbf{x}\|}. \tag{5.21}$$

Using (5.20) once more and recalling that $A\mathbf{x} = \mathbf{b}$, we finally obtain

$$\frac{\|\mathbf{x} - \widehat{\mathbf{x}}\|}{\|\mathbf{x}\|} \leq \frac{\lambda_{max}}{\lambda_{min}} \frac{\|\boldsymbol{\delta}\mathbf{b}\|}{\|\mathbf{b}\|} \tag{5.22}$$

We can conclude that the relative error in the solution depends on the relative error in the data through the following constant (≥ 1)

$$K(\mathrm{A}) = \frac{\lambda_{max}}{\lambda_{min}} \tag{5.23}$$

which is called *spectral condition number of the matrix* A. $K(\mathrm{A})$ can be
cond   computed in MATLAB using the command `cond(A)`. Other definitions for the condition number are available for non symmetric matrices, see [QSS00], Chapter 3.

Remark 5.2 The MATLAB command `cond(A)` allows the computation of the condition number of any type of matrix A, even those which are not sym-
condest   metric and positive definite. A special command `condest(A)` is available to
rcond   compute an approximation of the condition number of A, and one `rcond(A)` for its reciprocal, with a substantial saving of floating point operations. If the matrix A is ill-conditioned (*i.e.* $K(\mathrm{A}) \gg 1$), the computation of its condition number can be very inaccurate. Consider for instance the tridiagonal matrices $A_n = \mathrm{tridiag}(-1, 2, -1)$ for different values of n. A_n is symmetric and positive definite, its eigenvalues are $\lambda_j = 2 - 2\cos(j\theta)$, for $j = 1, \ldots, n$, with $\theta = \pi/(n+1)$, hence $K(A_n)$ can be computed exactly. In Figure 5.5 we report the value of the error $E_K(n) = |K(A_n) - \mathrm{cond}(A_n)|/K(A_n)$. Note that $E_K(n)$ increases when n increases.

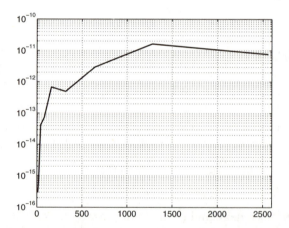

Fig. 5.5. Behavior of $E_K(n)$ as a function of n (in logarithmic scale)

A more involved proof would lead to the following more general result in the case where δA is an arbitrary symmetric and positive definite matrix "small enough" to satisfy $\lambda_{max}(\delta A) < \lambda_{min}(A)$:

$$\frac{\|\mathbf{x} - \widehat{\mathbf{x}}\|}{\|\mathbf{x}\|} \leq \frac{K(A)}{1 - \lambda_{max}(\delta A)/\lambda_{min}} \left(\frac{\lambda_{max}(\delta A)}{\lambda_{max}} + \frac{\|\delta \mathbf{b}\|}{\|\mathbf{b}\|} \right).$$

If $K(A)$ is "small", that is of the order of the unity, A is said to be *well conditioned*. In that case, small errors in the data will lead to errors of the same order of magnitude in the solution. This could not occur in the case of *ill conditioned* matrices.

Example 5.8 For the Hilbert matrix introduced in Example 5.7, $K(A_n)$ is a rapidly increasing function of n. One has $K(A_4) > 15000$, while if $n > 13$ the condition number is so high that MATLAB warns that the matrix is "close to singular". Actually, $K(A_n)$ grows at an exponential rate: $K(A_n) \simeq e^{3.5n}$ (see, [Hig96]). This provides an indirect explanation of the bad results obtained in Example 5.7.

Inequality (5.22) can be reformulated by the help of the *residual* $\mathbf{r}$:

$$\mathbf{r} = \mathbf{b} - A\widehat{\mathbf{x}}. \tag{5.24}$$

Should $\widehat{\mathbf{x}}$ be the exact solution, the residual would be the null vector. Thus, in general, $\mathbf{r}$ can be regarded as an *estimator* of the error $\mathbf{x} - \widehat{\mathbf{x}}$. The extent to which the residual is a good error estimator depends on the size of the condition number of A. Indeed, observing that $\delta \mathbf{b} = A(\widehat{\mathbf{x}} - \mathbf{x}) = A\widehat{\mathbf{x}} - \mathbf{b} = -\mathbf{r}$, we deduce from (5.22) that

$$\boxed{\frac{\|\mathbf{x} - \widehat{\mathbf{x}}\|}{\|\mathbf{x}\|} \leq K(A)\frac{\|\mathbf{r}\|}{\|\mathbf{b}\|}} \tag{5.25}$$

Thus if $K(A)$ is "small", we can be sure that the error is small provided that the residual is small, whereas this might not be true when $K(A)$ is "large".

Example 5.9 The residuals associated with the computed solution of the linear systems of Example 5.7 are very small (their norms vary between 10^{-16} and 10^{-11}); however the computed solutions differ remarkably from the exact solution.

See the Exercises 5.9-5.10.

5.4   How to solve a tridiagonal system

In many applications (see for instance Chapter 8), we have to solve a system whose matrix has the form

$$
A = \begin{bmatrix}
a_1 & c_1 & & 0 \\
e_2 & a_2 & \ddots & \\
& \ddots & \ddots & c_{n-1} \\
0 & & e_n & a_n
\end{bmatrix}.
$$

This matrix is called *tridiagonal* since the only elements that can be non-null belong to the main diagonal and to the first super and sub diagonals.

Assume that the matrix coefficients are real numbers. Then if the Gauss factorization LU of A exists, the factors L and U must be *bidiagonals* (lower and upper, respectively), more precisely:

$$
L = \begin{bmatrix}
1 & & & 0 \\
\beta_2 & 1 & & \\
& \ddots & \ddots & \\
0 & & \beta_n & 1
\end{bmatrix}, \quad
U = \begin{bmatrix}
\alpha_1 & c_1 & & 0 \\
& \alpha_2 & \ddots & \\
& & \ddots & c_{n-1} \\
0 & & & \alpha_n
\end{bmatrix}.
$$

The unknown coefficients α_i and β_i can be determined by requiring that the equality LU = A holds. This yields the following recursive relations for the computation of the L and U factors:

$$
\alpha_1 = a_1, \quad \beta_i = \frac{e_i}{\alpha_{i-1}}, \quad \alpha_i = a_i - \beta_i c_{i-1}, \; i = 2, \ldots, n. \tag{5.26}
$$

Using (5.26), we can easily solve the two bidiagonal systems $Ly = b$ and $Ux = y$, to obtain the following formulae:

$$
(Ly = b) \quad y_1 = b_1, \quad y_i = b_i - \beta_i y_{i-1}, \quad i = 2, \ldots, n, \tag{5.27}
$$

$$
(Ux = y) \; x_n = \frac{y_n}{\alpha_n}, \quad x_i = (y_i - c_i x_{i+1})/\alpha_i, \quad i = n-1, \ldots, 1. \tag{5.28}
$$

This is known as the *Thomas algorithm* and allows the solution of the original system with a computational cost of the order of n operations.

spdiags   The MATLAB command spdiags allows the construction of a tridiagonal matrix. For instance, the commands

```
>> b=ones(10,1); a=2*b; c=3*b;
>> T=spdiags([b a c],-1:1,10,10);
```

compute the tridiagonal matrix $T \in \mathbb{R}^{10 \times 10}$ with elements equal to 2 on the main diagonal, 1 on the first subdiagonal and 3 on the first super-diagonal.

Note that T is stored in a *sparse mode*, according to which the only elements stored are those different than 0. When a system is solved by invoking the command \, MATLAB is able to recognize the type of matrix (in particular, whether it has been generated in a sparse mode) and select the most appropriate solution algorithm. In particular, when A is a tridiagonal matrix generated in sparse mode, the Thomas algorithm is the selected algorithm.

Example 5.10 Let us solve the linear system $Tx = b$ where T is a tridiagonal matrix constructed as previously with a variable size n, and b is chosen such that the exact solution is $x = (1, \ldots, 1)^T$. With the following instructions we solve this system (by invoking the \ command) for increasing values of n and compute the number of *ops* divided by n. This ratio for n very large tends to a constant, proving that the number of operations scales linearly with n, as expected.

```
>> ratio = [ ];
>> for k = 50:25:200
     b=ones(k,1); a=2*b; c=3*b;
     T=spdiags([b a c],-1:1,k,k);
     rhs = T*ones(k,1);
     flops(0); x=T\rhs; ratio = [ratio, flops/k];
end
>> ratio =
  Columns 1 through 7
   24.5000   24.6133   24.7300   24.7520   24.8067   24.7771   24.8150
```

Let us summarize

1. The factorization LU of A consists of computing a lower triangular matrix L and an upper triangular matrix U such that A = LU;

2. the factorization LU, provided it exists, is not unique. However, it can be determined unequivocally by furnishing an additional condition such as, *e.g.*, setting the diagonal elements of L equal to 1. This is called Gauss factorization;

3. the Gauss factorization exists if and only if the principal subma-trices of A of order $1, 2, \ldots, (n-1)$ are non singular (otherwise

at least one *pivot* element is null); further, it is unique provided $\det(A) \neq 0$;

4. if a null *pivot* element is generated, a new *pivot* element can be obtained by exchanging in a suitable manner two rows (or columns) of our system. This is the *pivoting* strategy;

5. the computation of the Gauss factorization requires about $2n^3/3$ operations, and only an order of n operations in the case of tridiagonal systems;

6. for symmetric and positive definite matrices we can use the Cholesky factorization $A = HH^T$, where H is a lower triangular matrix, and the computational cost is of the order of $n^3/3$ operations;

7. the sensitivity of the result to perturbation of data depends on the condition number of the system matrix; more precisely, the accuracy of the computed solution can be low for ill conditioned matrices.

5.5 Iterative methods

An iterative method for the solution of the linear system (5.1) consists of setting up a sequence of vectors $\mathbf{x}^{(k)} \in \mathbb{R}^n$ that *converge* to the exact solution $\mathbf{x}$, that is

$$\lim_{k \to \infty} \mathbf{x}^{(k)} = \mathbf{x}, \tag{5.29}$$

for any given initial vector $\mathbf{x}^{(0)} \in \mathbb{R}^n$. A possible strategy able to realize this process can be based on the following recursive definition

$$\mathbf{x}^{(k+1)} = B\mathbf{x}^{(k)} + \mathbf{g}, \qquad k \geq 0, \tag{5.30}$$

where B is a suitable matrix (depending on A) and $\mathbf{g}$ is a suitable vector (depending on A and $\mathbf{b}$), which must satisfy the relation

$$\mathbf{x} = B\mathbf{x} + \mathbf{g}. \tag{5.31}$$

Since $\mathbf{x} = A^{-1}\mathbf{b}$ this yields $\mathbf{g} = (I - B)A^{-1}\mathbf{b}$.

Let $\mathbf{e}^{(k)} = \mathbf{x} - \mathbf{x}^{(k)}$ define the error at the k-th step. By subtracting (5.30) from (5.31), we obtain

$$\mathbf{e}^{(k+1)} = B\mathbf{e}^{(k)}.$$

For this reason B is called the *iteration matrix* associated with (5.30). If B is symmetric and positive definite, by (5.20) we have

$$\|\mathbf{e}^{(k+1)}\| = \|\mathbf{B}\mathbf{e}^{(k)}\| \le \rho(\mathbf{B})\|\mathbf{e}^{(k)}\|, \qquad \forall k \ge 0.$$

We have denoted by $\rho(\mathbf{B})$ the *spectral radius* of B, that is, the maximum modulus of eigenvalues of B. By iterating the same inequality backward, we obtain

$$\|\mathbf{e}^{(k)}\| \le [\rho(\mathbf{B})]^k \|\mathbf{e}^{(0)}\|, \quad k \ge 0. \tag{5.32}$$

Thus $\mathbf{e}^{(k)} \to \mathbf{0}$ as $k \to \infty$ for every possible $\mathbf{e}^{(0)}$ (and henceforth $\mathbf{x}^{(0)}$) provided that $\rho(\mathbf{B}) < 1$. Actually, this property is also necessary for convergence.

Should, by any chance, an approximate value of $\rho(\mathbf{B})$ be available, (5.32) would allow us to deduce the minimum number of iterations k_{min} that are needed to damp the initial error by a factor ε. Indeed, k_{min} would be the lowest positive integer for which $[\rho(\mathbf{B})]^{k_{min}} \le \varepsilon$.

In conclusion, the following result holds:

Proposition 5.2 *For an iterative method of the form* (5.30) *whose iteration matrix satisfies* (5.31), *convergence for any* $\mathbf{x}^{(0)}$ *holds iff* $\rho(\mathbf{B}) < 1$. *Moreover, the smaller* $\rho(\mathbf{B})$, *the fewer the number of iterations necessary to reduce the initial error by a given factor.*

5.5.1 How to construct an iterative method

A general technique to devise an iterative method is based on a *splitting* of the matrix A, $\mathbf{A} = \mathbf{P} - (\mathbf{P} - \mathbf{A})$, being P a suitable nonsingular matrix (called the *preconditioner* of A). Then

$$\mathbf{x} = \mathbf{P}^{-1}(\mathbf{P} - \mathbf{A})\mathbf{x} + \mathbf{P}^{-1}\mathbf{b},$$

which has the form (5.31) provided that we set $\mathbf{B} = \mathbf{P}^{-1}(\mathbf{P} - \mathbf{A}) = \mathbf{I} - \mathbf{P}^{-1}\mathbf{A}$ and $\mathbf{g} = \mathbf{P}^{-1}\mathbf{b}$. Correspondingly, we can define the following iterative method:

$$\mathbf{P}(\mathbf{x}^{(k+1)} - \mathbf{x}^{(k)}) = \mathbf{r}^{(k)}, \qquad k \ge 0,$$

where $\mathbf{r}^{(k)} = \mathbf{b} - \mathbf{A}\mathbf{x}^{(k)}$ denotes the residual vector at the k-th iteration. A generalization of this iterative method is the following

$$\mathbf{P}(\mathbf{x}^{(k+1)} - \mathbf{x}^{(k)}) = \alpha_k \mathbf{r}^{(k)}, \qquad k \ge 0 \tag{5.33}$$

where $\alpha_k \neq 0$ is a parameter that may change at every iteration k.

The method (5.33) requires at each step the solution of the linear system

$$\mathbf{P}\mathbf{z}^{(k)} = \mathbf{r}^{(k)}, \qquad (5.34)$$

then the new iterate is defined by $\mathbf{x}^{(k+1)} = \mathbf{x}^{(k)} + \alpha_k \mathbf{z}^{(k)}$. For that reason the matrix P ought to be chosen in such a way that the computational cost for the solution of (5.34) be quite low (*e.g.*, every P either diagonal or triangular or tridiagonal will serve the purpose). Let us now consider some special instance of iterative methods which take the form (5.33).

The Jacobi method

If the diagonal entries of A are nonzero, we can set $\mathbf{P} = \mathbf{D}$, where D is the diagonal matrix of the diagonal entries of A. The Jacobi method corresponds to this choice with the assumption $\alpha_k = 1$ for all k. Then from (5.33) we obtain

$$\mathbf{D}\mathbf{x}^{(k+1)} = \mathbf{b} - (\mathbf{A} - \mathbf{D})\mathbf{x}^{(k)}, \qquad k \geq 0,$$

or, componentwise,

$$x_i^{(k+1)} = \frac{1}{a_{ii}} \left(b_i - \sum_{j=1, j\neq i}^{n} a_{ij} x_j^{(k)} \right), \quad i = 1, \ldots, n \qquad (5.35)$$

where $k \geq 0$ and $\mathbf{x}^{(0)} = (x_1^{(0)}, x_2^{(0)}, \ldots, x_n^{(0)})^T$ is the initial vector.

The iteration matrix is therefore

$$\mathbf{B} = \mathbf{D}^{-1}(\mathbf{D} - \mathbf{A}) = \begin{bmatrix} 0 & -a_{12}/a_{11} & \cdots & -a_{1n}/a_{11} \\ -a_{21}/a_{22} & 0 & & -a_{2n}/a_{22} \\ \vdots & & \ddots & \vdots \\ -a_{n1}/a_{nn} & -a_{n2}/a_{nn} & \cdots & 0 \end{bmatrix} \qquad (5.36)$$

The following result allows the verification of Proposition 5.2 without explicitly computing $\rho(\mathbf{B})$:

Proposition 5.3 *If the matrix A is strictly diagonally dominant by row, then the Jacobi method converges.*

As a matter of fact, we can verify that $\rho(B) < 1$, where B is given in (5.36). To start with, we note that the diagonal elements of A are nonnull owing to the strict diagonal dominance. Let λ be a generic eigenvalue of B and $\mathbf{x}$ an associated eigenvector. Then

$$\sum_{j=1}^{n} b_{ij} x_j = \lambda x_i, \quad i = 1, \ldots, n.$$

Assume for simplicity that $\max_{k=1,\ldots,n} |x_k| = 1$ (this is not restrictive since an eigenvector is defined up to a multiplicative constant) and let x_i be the component whose modulus is equal to 1. Then

$$|\lambda| = \left| \sum_{j=1}^{n} b_{ij} x_j \right| = \left| \sum_{j=1, j \neq i}^{n} b_{ij} x_j \right| \leq \sum_{j=1, j \neq i}^{n} \left| \frac{a_{ij}}{a_{ii}} \right|,$$

having noticed that B has only null diagonal elements. Therefore $|\lambda| < 1$ thanks to the assumption made on A.

The Jacobi method is implemented in the Program 8 setting in the input parameter P='J'. Input parameters are: the system matrix A, the right hand side b, the initial vector x0 and the maximum number of iterations allotted, nmax. The iterative procedure is terminated as soon as the ratio between the Euclidean norm of the current residual and that of the initial residual is less than a prescribed tolerance tol (for a justification of this stopping criterion, see Section 5.6).

Program 8 - itermeth: General iterative method

```
function [x, iter]= itermeth(A,b,x0,nmax,tol,P)
%ITERMETH    General iterative method
%   X = ITERMETH(A,B,X0,NMAX,TOL,P) attempts to solve the system of
%   linear equations A*X=B for X. The N-by-N coefficient matrix A
%   must be not singular and the right hand side column vector B must
%   have length N. If P='J' the Jacobi method is used, if P='G' the
%   Gauss-Seidel method is selected. Otherwise, P is a N-by-N matrix
%   that play the role of a preconditioner. TOL specifies the tolerance
%   of the method. NMAX specifies the maximum number of iterations.
[n,n]=size(A);
if nargin == 6
  if ischar(P)==1
    if P=='J'
      L = diag(diag(A));
      U = eye(n); beta = 1; alpha = 1;
    elseif P == 'G'
      L = tril(A);
```

```
    U = eye(n); beta = 1; alpha = 1;
    end
  else
    [L,U]=lu(P); beta = 0;
  end
else
  L = eye(n); U = L; beta = 0;
end
iter = 0;
r = b - A * x0;
r0 = norm(r);
err = norm (r); x = x0;
while err > tol & iter < nmax
  iter = iter + 1;
  z = L\r;
  z = U\z;
  if beta == 0
    alpha = z'*r/(z'*A*z);
  end
  x = x + alpha*z;
  r = b - A * x;
  err = norm (r) / r0;
end
```

The Gauss-Seidel method

When applying the Jacobi method, each individual component of the new vector, say $x_i^{(k+1)}$, is computed independently of the others. This may suggest that a faster convergence could be (hopefully) achieved if the new components already available $x_j^{(k+1)}$, $j = 1, \ldots, i - 1$, together with the old ones $x_j^{(k)}$, $j \geq i$, are used for the calculation of $x_i^{(k+1)}$. This would lead to modifying (5.35) as follows: for $k \geq 0$ (still assuming that $a_{ii} \neq 0$ for $i = 1, \ldots, n$)

$$
x_i^{(k+1)} = \frac{1}{a_{ii}} \left(b_i - \sum_{j=1}^{i-1} a_{ij} x_j^{(k+1)} - \sum_{j=i+1}^{n} a_{ij} x_j^{(k)} \right) , i = 1, .., n \quad (5.37)
$$

The updating of the components is made in *sequential* mode, whereas in the original Jacobi method it is made *simultaneously* (or in parallel). The new method, which is called the *Gauss-Seidel method*, corresponds to the choice $P = D - E$ and $\alpha_k = 1$, $k \geq 0$, in (5.33), where E is a lower triangular matrix whose non null entries are $e_{ij} = -a_{ij}$, $i = 2, \ldots, n$,

$j = 1, \ldots, i - 1$. The corresponding iteration matrix is then

$$B = (D - E)^{-1}(P - A).$$

A possible generalization is the so-called *relaxation method* in which $P = \frac{1}{\omega}D - E$, where $\omega \neq 0$ is the relaxation parameter, and $\alpha_k = 1$, $k \geq 0$ (see Exercise 5.13).

Also for Gauss-Seidel method there exist special matrices A whose associated iteration matrices satisfy the assumptions of Proposition 5.2 (those guaranteeing convergence). Among them let us mention:

1. matrices strictly diagonally dominant by row;

2. matrices which are symmetric and positive definite.

The Gauss-Seidel method is implemented in the Program 8 setting the input parameter P equal to 'G'.

There are no general results stating that the Gauss-Seidel method converges faster than Jacobi's. However, in some special instances this is the case, as stated by the following proposition:

Proposition 5.4 *Let* A *be a tridiagonal* $n \times n$ *nonsingular matrix whose diagonal elements are all nonnull. Then the Jacobi method and the Gauss-Seidel method are either both divergent or both convergent. In the latter case, the Gauss-Seidel method is faster than Jacobi's; more precisely the spectral radius of its iteration matrix is equal to the square of that of Jacobi.*

Example 5.11 Let us consider a linear system $Ax = b$ where b is chosen in such a way that the solution is the unit vector $(1, 1, \ldots, 1)^T$ and A is the 10×10 tridiagonal matrix whose diagonal entries are all equal to 3, the entries of the first lower diagonal are equal to -2 and those of the upper diagonal are all equal to -1. Both Jacobi and Gauss-Seidel methods converge since the spectral radii of their iteration matrices are strictly less than 1. More precisely, by starting from a null initial vector and setting `tol` $=10^{-12}$, the Jacobi method converges in 277 iterations while only 143 iterations are requested from Gauss-Seidel's. To get this result we have used the following instructions:

```
>> n=10; A = 3*eye(n) - 2*diag(ones(n-1,1),1) - diag(ones(n-1,1),-1);
>> b=A*ones(n,1);
>> [x,iter]=itermeth(A,b,zeros(n,1),400,1.e-12,'J'); iter
iter =
  277
>> [x,iter]=itermeth(A,b,zeros(n,1),400,1.e-12,'G'); iter
iter =
  143
```

See the Exercises 5.11-5.14.

5.6   When should an iterative method be stopped?

In theory iterative methods require an infinite number of iterations to converge to the exact solution. In practice, this is neither reasonable nor necessary. Indeed we do not really need to achieve the exact solution, but rather an approximation $\mathbf{x}^{(k)}$ for which we can guarantee that the error be lower than a desired tolerance ϵ. On the other hand, since the error is itself unknown (as it depends on the exact solution), we need a suitable *a posteriori* error estimator which predicts the error starting from quantities that have already been computed.

The first type of estimator is represented by the *residual* which is defined as

$$\mathbf{r}^{(k)} = \mathbf{b} - \mathrm{A}\mathbf{x}^{(k)},$$

being $\mathbf{x}^{(k)}$ the approximate value of the solution at the k-th iteration.

More precisely, we could stop our iterative method at the first iteration step k_{min} for which

$$\|\mathbf{r}^{(k_{min})}\| \le \varepsilon \|\mathbf{b}\|.$$

Setting $\widehat{\mathbf{x}} = \mathbf{x}^{(k_{min})}$ and $\mathbf{r} = \mathbf{r}^{(k_{min})}$ in (5.25) we would obtain

$$\frac{\|\mathbf{e}^{(k_{min})}\|}{\|\mathbf{x}\|} \le \varepsilon K(\mathrm{A}),$$

which is an estimate for the relative error. We deduce that the control on the residual is meaningful only for those matrices whose condition number is reasonably small.

Example 5.12 Let us consider the linear system (5.1) where A=A$_{20}$ is the Hilbert matrix of dimension 20 introduced in Example 5.7 and **b** is constructed in such a way that the exact solution is $\mathbf{x} = (1, 1, \ldots, 1)^T$. Since A is symmetric and positive definite the Gauss-Seidel method surely converges. We use Program 8 to solve this system taking x0 to be the null initial vector and setting a tolerance on the residual equal to 10^{-5}. The method converges in 472 iterations; however the relative error is very large and equals 0.26. This is due to the fact that A is extremely ill conditioned, having $K(\mathrm{A}) \simeq 10^{17}$. In Figure 5.6 we show the behavior of the residual (normalized to the initial one) and that of the error as the number of iterations increases.

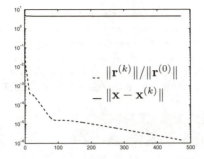

Fig. 5.6. Behavior of residual (dashed line) and error (solid line) for Gauss-Seidel iterations applied to the system of Example 5.12

An alternative approach is based on the use of a different error estimator, namely the *increment* $\boldsymbol{\delta}^{(k_{min})} = \mathbf{x}^{(k_{min}+1)} - \mathbf{x}^{(k_{min})}$. More precisely, we can stop our iterative method at the first iteration step k_{min} for which

$$\|\boldsymbol{\delta}^{(k_{min})}\| \leq \varepsilon \|\mathbf{b}\|.$$

In the special case in which B is symmetric and positive definite, we have

$$\|\mathbf{e}^{(k)}\| = \|\mathbf{e}^{(k+1)} - \boldsymbol{\delta}^{(k)}\| \leq \rho(\mathrm{B})\|\mathbf{e}^{(k)}\| + \|\boldsymbol{\delta}^{(k)}\|.$$

Since $\rho(\mathrm{B})$ should be less than 1 in order for the method to converge, we deduce

$$\|\mathbf{e}^{(k)}\| \leq \frac{1}{1 - \rho(\mathrm{B})}\|\boldsymbol{\delta}^{(k)}\| \tag{5.38}$$

From the last inequality we see that the control on the increment is meaningful only if $\rho(\mathrm{B})$ is much smaller than 1 since in that case the error will be of the same size as the increment.

In fact, the same conclusion holds even if B is not symmetric and positive definite (as it occurs for the Jacobi and Gauss-Seidel methods); however in that case (5.38) is no longer true.

Example 5.13 Let us consider a system whose matrix $A \in \mathbb{R}^{50 \times 50}$ is tridiagonal and symmetric with entries equal to 2.001 on the main diagonal and equal to 1 on the two other diagonals. As usual, the right hand side $\mathbf{b}$ is chosen in such a way that the unit vector $(1, \ldots, 1)^T$ is the exact solution. Since A is tridiagonal with strict diagonal dominance, the Gauss-Seidel method will converge about twice as fast as the Jacobi method (in view of Proposition 5.4). Let us use Program 8 to solve our system in which we replace the stopping criterion based on the residual by that based on the increment. Using a null initial vector and setting the tolerance tol$= 10^{-5}$, after 1604 iterations the program returns a solution whose error 0.0029 is quite large. The reason is that the spectral radius of the iteration matrix is equal to 0.9952, which is very close to 1. Should the diagonal entries be set equal to 3, after only 17 iterations we would have obtained an error equal to 10^{-5}. In fact in that case the spectral radius of the iteration matrix would be equal to 0.428.

5.7   Richardson method

Let us now consider methods (5.33) for which the acceleration parameters α_k are non-null. We call *stationary* the case when $\alpha_k = \alpha$ (a given constant) for any $k \geq 0$, *dynamic* the case in which α_k may change along the iterations. In this framework the non singular matrix P is still called a *preconditioner* of A.

The crucial issue is the way the parameters are chosen. In this respect, the following result holds (see, *e.g.*, [QV94, Chap.2], [Axe94]).

Proposition 5.5 *If both* P *and* A *are symmetric and positive definite, the stationary Richardson method converges for every possible choice of* $\mathbf{x}^{(0)}$ *iff* $0 < \alpha < 2/\lambda_{max}$, *where* $\lambda_{max}(> 0)$ *is the maximum eigenvalue of* $P^{-1}A$. *Moreover, the spectral radius* $\rho(B_\alpha)$ *of the iteration matrix* $B_\alpha = I - \alpha P^{-1}A$ *is least when* $\alpha = \alpha_{opt}$, *where*

$$
\boxed{\alpha_{opt} = \frac{2}{\lambda_{min} + \lambda_{max}}}
\tag{5.39}
$$

λ_{min} *being the minimum eigenvalue of* $P^{-1}A$.

Under the same assumption on P *and* A, *the dynamic Richardson method converges if for instance* α_k *is chosen in the following way:*

$$
\boxed{\alpha_k = \frac{(\mathbf{z}^{(k)})^T \mathbf{r}^{(k)}}{(\mathbf{z}^{(k)})^T A \mathbf{z}^{(k)}} \qquad \forall k \geq 0}
$$

where $\mathbf{z}^{(k)} = P^{-1}\mathbf{r}^{(k)}$ *is the preconditioned residual. The method (5.33) with this choice of* α_k *is called the preconditioned gradient method, or simply the gradient method when the preconditioner* P *is the identity matrix.*

In both cases, the following convergence estimate holds:

$$
\|\mathbf{e}^{(k)}\|_A \leq \left(\frac{K(P^{-1}A) - 1}{K(P^{-1}A) + 1} \right)^k \|\mathbf{e}^{(0)}\|_A, \quad k \geq 0,
\tag{5.40}
$$

where $\|\mathbf{v}\|_A = \sqrt{\mathbf{v}^T A \mathbf{v}}$, $\forall \mathbf{v} \in \mathbb{R}^n$, *is the so-called energy norm associated with the matrix* A.

The dynamic version should therefore be preferred to the stationary one since it does not require the knowledge of the extreme eigenvalues of $P^{-1}A$. Rather, the parameter α_k is determined in terms of quantities which are already available from the previous iteration.

We can rewrite the preconditioned gradient method more efficiently through the following algorithm (derivation is left as an exercise): given $\mathbf{x}^{(0)}$, for every $k \geq 0$ do

$$
\begin{aligned}
& P\mathbf{z}^{(k)} = \mathbf{r}^{(k)}, \\
& \alpha_k = \frac{(\mathbf{z}^{(k)})^T \mathbf{r}^{(k)}}{(\mathbf{z}^{(k)})^T A\mathbf{z}^{(k)}}, \\
& \mathbf{x}^{(k+1)} = \mathbf{x}^{(k)} + \alpha_k \mathbf{z}^{(k)}, \\
& \mathbf{r}^{(k+1)} = \mathbf{r}^{(k)} - \alpha_k A\mathbf{z}^{(k)}
\end{aligned}
\tag{5.41}
$$

The same algorithm can be used to implement the stationary Richardson method by simply replacing α_k with the constant value α.

From (5.40), we deduce that if $P^{-1}A$ is ill conditioned the convergence rate will be very low even for $\alpha = \alpha_{opt}$ (as in that case $\rho(B_{\alpha_{opt}}) \simeq 1$). This circumstance can be avoided provided that a convenient choice of P is made. This is the reason why P is called the preconditioner or the preconditioning matrix.

If A is a generic matrix it may be a difficult task to find a preconditioner which guarantees an optimal trade-off between damping the condition number and keeping the computational cost for the solution of the system (5.34) reasonably low.

The dynamic Richardson method is implemented in Program 8 where the input parameter P stands for the preconditioning matrix (when not prescribed, the program implements the unpreconditioned method by setting P=I).

Example 5.14 This example, of theoretical interest only, has the purpose of comparing the convergence behavior of Jacobi, Gauss-Seidel and gradient methods applied to solve the following (micro) linear system:

$$
2x_1 + x_2 = 1, \quad x_1 + 3x_2 = 0
\tag{5.42}
$$

with initial vector $\mathbf{x}^{(0)} = (1, 1/2)^T$. Note that the system matrix is symmetric and positive definite, and that the exact solution is $\mathbf{x} = (3/5, -1/5)^T$. We report in Figure 5.7 the behavior of the relative residual $E^{(k)} = \|\mathbf{r}^{(k)}\| / \|\mathbf{r}^{(0)}\|$ for the three methods above. Iterations are stopped at the first iteration k_{min} for which $E^{(k_{min})} \leq 10^{-14}$. The gradient method appears to converge the fastest.

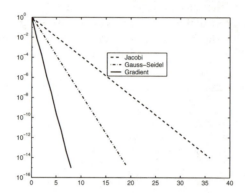

Fig. 5.7. Convergence history for Jacobi, Gauss-Seidel and gradient methods applied to system (5.42)

Example 5.15 Let us consider a system $A\mathbf{x} = \mathbf{b}$ where $A \in \mathbb{R}^{100 \times 100}$ is a pentadiagonal matrix whose main diagonal has all entries equal to 4, while the first and third lower and upper diagonals have all entries equal to -1. As customary, $\mathbf{b}$ is chosen in such a way that $\mathbf{x} = (1, \dots, 1)^T$ is the exact solution of our system. Let us use Program 8 which implements the preconditioned Richardson method. We fix `tol=1.e-05`, `nmax=5000`, `x0=zeros(100,1)` and the preconditioner P is a tridiagonal matrix whose diagonal elements are all equal to 2, while the elements on the lower and upper diagonal are all equal to -1. Both A and P are symmetric and positive definite. The method converges in 18 iterations, whereas Program 8 (implementing the Gauss-Seidel method) requires as many as 2421 iterations before fulfilling the same stopping criterion.

Example 5.16 (Direct and iterative methods) Let us go back to Example 5.7 on the Hilbert matrix and solve the system (for different values of n) by the preconditioned gradient method using the diagonal preconditioner D, made of the diagonal entries of the Hilbert matrix. We report also the results that would be obtained using a more efficient iterative method, the conjugate gradient method (CG). The CG method will be addressed in Section 5.8. We fix $\mathbf{x}^{(0)}$ to be the null vector and iterate untill the relative residual is less than 10^{-6}. In Table 5.1 we report the absolute errors (with respect to the exact solution) that are obtained with the above iterative methods and those that would be obtained by using the Gauss LU factorization approach. The error degenerates when n gets large and the direct method is used. On the other hand, we can appreciate the benefical effect that a suitable iterative method as the CG scheme, can have on the number of iterations.

n	$K(A)$	LU Error	PG Error	Iter.	PCG Error	Iter.
4	1.55e+04	3.90e-13	1.74-02	995	2.24e-02	3
6	1.50e+07	2.62e-10	8.80e-03	1813	9.50e-03	9
8	1.53e+10	1.34e-07	1.78e-02	1089	2.13e-02	4
10	1.60e+13	7.60e-04	2.52e-03	875	6.98e-03	5
12	1.67e+16	4.97e-01	1.76e-02	1355	1.12e-02	5
14	3.04e+17	6.65e+00	1.46e-02	1379	1.61e-02	5

Tab. 5.1. Errors obtained using the preconditioned gradient method (PG), the preconditioned conjugate gradient method (PCG) and the Gauss LU factorization method for the solution of the Hilbert system

See the Exercises 5.15-5.17.

Let us summarize

1. An iterative method for the solution of a linear system starts from a given initial vector $\mathbf{x}^{(0)}$ and builds up a sequence of vectors $\mathbf{x}^{(k)}$ which we require to converge to the exact solution as $k \to \infty$;

2. an iterative method converges for every possible choice of the initial vector $\mathbf{x}^{(0)}$ iff the spectral radius of the iteration matrix is strictly less than 1;

3. classical iterative methods are those of Jacobi and Gauss-Seidel. A sufficient condition for convergence is that the system matrix be strictly diagonally dominant by row (or symmetric and definite positive in the case of Gauss-Seidel);

4. in the Richardson method convergence is accelerated thanks to the introduction of a parameter and (possibly) a convenient preconditioning matrix;

5. there are two possible stopping criteria for an iterative method: controlling the residual or controlling the increment. The former is meaningful if the system matrix is well conditioned, the latter if the spectral radius of the iteration matrix is not close to 1.

5.8   What we haven't told you

Several efficient variants of the Gauss LU factorization are available for sparse systems of large dimension. Among the most advanced, we quote the so-called *multifrontal method* which makes use of a suitable reordering of the system unknowns in order to keep the triangular factors L and U as sparse as possible. More on this issue is available on [GL89], [DD95] and [Iro70].

Concerning iterative methods, there exists a broad family of modern methods, the so-called *Krylov methods*, which are more efficient than those presented above. Some of them feature the notable property of finite termination, that is, in exact arithmetic they provide the exact solution in a finite number of iterations. We can mention the *conjugate gradient method* (which can be applied if the system matrix is symmetric and positive definite) and the *GMRES* (Generalized Minimum RESidual). Their description is provided, *e.g.*, in [Axe94] and [Saa96]. They are available in the MATLAB toolbox `sparfun` under the name of `pcg` and `gmres`.

pcg
gmres

As already pointed out, iterative methods converge slowly if the system matrix is severely ill conditioned. Several preconditioning strategies have been developed (see, *e.g.*, [dV89]). Some of them are purely algebraic, that is, they are based on incomplete (or inexact) factorizations of the given system matrix, and are implemented in the MATLAB functions `luinc` or `cholinc`. Other strategies are developed *ad hoc* by exploiting the meaning and the structure of the problem which has generated the linear system at hand.

luinc
cholinc

Finally it is worthwhile mentioning the *multigrid* algorithms which are based on the sequential use of a hierarchy of systems of variable dimensions that "resemble" the original one, allowing a clever strategy of reduction of the error (see, *e.g.*, [Hac85], [Wes91] and [Hac94]).

5.9   Exercises

Exercise 5.1 For a given matrix $A \in \mathbb{R}^{n \times n}$ find the number of operations (as a function of n) that are needed for computing its determinant by the recursive formula (1.8).

magic

Exercise 5.2 Use the MATLAB command `magic(n)`, n=3, 4, ..., 500, to construct the magic squares of order n, that is, those matrices having entries for which the sum of the elements by rows, columns or diagonals are identical. Then compute their determinants by the command `det` introduced in Section

1.3 and the CPU time that is needed for this computation using the command `cputime`. Finally, approximate these data by the least-square method and deduce that the CPU time scales approximately as n^3.

Exercise 5.3 Find for which values of ε the matrix defined in (5.12) does not satisfy the hypotheses of Proposition 5.1. For which value of ε does this matrix become singular? Is it possible to compute the LU factorization in that case?

Exercise 5.4 Verify that the number of operations necessary to compute the LU factorization of a square matrix A of dimension n is approximately $2n^3/3$.

Exercise 5.5 Show that the LU factorization of A can be used for computing the inverse matrix A^{-1}. (Observe that the j-th column vector of A^{-1} satisfies the linear system $A\mathbf{y}_j = \mathbf{e}_j$, $\mathbf{e}_j$ being the vector whose components are all null except the j-th component which is 1.)

Exercise 5.6 Compute the factors L and U of the matrix of Example 5.6 and verify that the LU factorization is inaccurate.

Exercise 5.7 Explain why partial pivoting by row is not convenient for symmetric matrices.

Exercise 5.8 Consider the linear system $A\mathbf{x} = \mathbf{b}$ with

$$A = \begin{bmatrix} 2 & -2 & 0 \\ \varepsilon - 2 & 2 & 0 \\ 0 & -1 & 3 \end{bmatrix},$$

and $\mathbf{b}$ such that the corresponding solution is $\mathbf{x} = (1,1,1)^T$ and ε is a positive real number. Compute the Gauss factorization of A and note that $l_{32} \to \infty$ when $\varepsilon \to 0$. In spite of that, verify that the computed solution is accurate.

Exercise 5.9 Consider the linear systems $A_i\mathbf{x}_i = \mathbf{b}_i$, $i = 1, 2, 3$, with

$$A_1 = \begin{bmatrix} 15 & 6 & 8 & 11 \\ 6 & 6 & 5 & 3 \\ 8 & 5 & 7 & 6 \\ 11 & 3 & 6 & 9 \end{bmatrix}, \quad A_i = (A_1)^i, \ i = 2, 3,$$

and $\mathbf{b}_i$ such that the solution is always $\mathbf{x}_i = (1,1,1,1)^T$. Solve the system by the Gauss factorization using partial pivoting by row, and comment on the obtained results.

Exercise 5.10 Show that for a symmetric and positive definite matrix A we have $K(A^2) = (K(A))^2$.

Exercise 5.11 Analyse the convergence properties of the Jacobi and Gauss-Seidel methods for the solution of a linear system whose matrix is

$$A = \begin{bmatrix} \alpha & 0 & 1 \\ 0 & \alpha & 0 \\ 1 & 0 & \alpha \end{bmatrix}, \qquad \alpha \in \mathbb{R}.$$

Exercise 5.12 Provide a sufficient condition on β so that both the Jacobi and Gauss-Seidel methods converge when applied for the solution of a system whose matrix is

$$A = \begin{bmatrix} -10 & 2 \\ \beta & 5 \end{bmatrix}. \tag{5.43}$$

Exercise 5.13 For the solution of the linear system $A\mathbf{x} = \mathbf{b}$ with $A \in \mathbb{R}^{n \times n}$, consider the *relaxation method*: given $\mathbf{x}^{(0)} = (x_1^{(0)}, \ldots, x_n^{(0)})^T$, for $k = 0, 1, \ldots$ compute

$$r_i^{(k)} = b_i - \sum_{j=1}^{i-1} a_{ij} x_j^{(k+1)} - \sum_{j=i}^{n} a_{ij} x_j^{(k)}, \quad x_i^{(k+1)} = (1-\omega) x_i^{(k)} + \omega \frac{r_i^{(k)}}{a_{ii}},$$

for $i = 1, \ldots, n$, where ω is a real parameter. Find the explicit form of the corresponding iterative matrix, then verify that the condition $0 < \omega < 2$ is necessary for the convergence of this method. Note that if $\omega = 1$ this method reduces to the Gauss-Seidel method. If $1 < \omega < 2$ the method is known as *SOR (successive over-relaxation)*.

Exercise 5.14 Consider the linear system $A\mathbf{x} = \mathbf{b}$ with $A = \begin{bmatrix} 3 & 2 \\ 2 & 6 \end{bmatrix}$ and say whether the Gauss-Seidel method converges, without explicitly computing the spectral radius of the iteration matrix.

Exercise 5.15 Compute the first iteration of the Jacobi, Gauss-Seidel and preconditioned gradient method (with preconditioner given by the diagonal of A) for the solution of system (5.42).

Exercise 5.16 Prove (5.39), then show that

$$\rho(B_{\alpha_{opt}}) = \frac{\lambda_{max} - \lambda_{min}}{\lambda_{max} + \lambda_{min}} = \frac{K(P^{-1}A) - 1}{K(P^{-1}A) + 1} \tag{5.44}$$

Exercise 5.17 Let us consider a set of $n = 20$ factories which produce 20 different goods. With reference to the Leontief model introduced in Problem 5.3, suppose that the matrix C has the following integer entries: $c_{ij} = i + j - 1$ for $i, j = 1, \ldots, n$, while $b_i = i$, for $i = 1, \ldots, 20$. Is it possible to solve this system by the gradient method? Propose a method based on the gradient method noting that, if A is non-singular, the matrix $A^T A$ is symmetric and positive definite.

6. Eigenvalues and eigenvectors

We consider the following problem: given a square matrix $A \in \mathbb{R}^{n \times n}$, find a scalar λ (real or complex) and a non-null vector $\mathbf{x}$ such that

$$A\mathbf{x} = \lambda\mathbf{x}. \tag{6.1}$$

Any such λ is called an *eigenvalue* of A, while $\mathbf{x}$ is the associated *eigenvector*. The latter is not unique; indeed all its multiples $\alpha\mathbf{x}$ with $\alpha \neq 0$, real or complex, are also eigenvectors associated with λ. Should $\mathbf{x}$ be available, λ can be recovered by using the *Rayleigh quotient* $\mathbf{x}^* A\mathbf{x}/\|\mathbf{x}\|^2$, $\mathbf{x}^*$ being the vector whose i-th component is equal to $\bar{x}_i$.

A number λ is an eigenvalue of A if it is a root of the following polynomial of degree n (called the *characteristic polynomial* of A):

$$p_A(\lambda) = \det(A - \lambda I).$$

Consequently, a square matrix of dimension n has exactly n eigenvalues (real or complex), not necessarily distinct. Also, if A has real entries, $p_A(\lambda)$ has real coefficients, and therefore complex eigenvalues of A necessarily occur in complex conjugate pairs.

Problem 6.1 (Dynamics) Consider the system of Figure 6.1 made of two pointwise bodies P_1 and P_2 of mass m, connected by 2 springs and free to move along the line joining P_1 and P_2. Let $x_i(t)$ denote the position occupied by P_i at time t for $i = 1, 2$. Then from the second law of dynamics we obtain

$$m\,\ddot{x}_1 = K(x_2 - x_1) - Kx_1, \quad m\,\ddot{x}_2 = K(x_1 - x_2).$$

We are interested in forced oscillations whose corresponding solution is $x_i = a_i \sin(\omega t + \phi)$, $i = 1, 2$, with $a_i \neq 0$. In this case we find that

$$\begin{aligned}
-ma_1\omega^2 &= K(a_2 - a_1) - Ka_1, \\
-ma_2\omega^2 &= K(a_1 - a_2).
\end{aligned} \tag{6.2}$$

This is a 2×2 homogeneous system which has a non-trivial solution a_1, a_2 iff the number $\lambda = m\omega^2/K$ is an eigenvalue of the matrix

$$A = \begin{bmatrix} 2 & -1 \\ -1 & 1 \end{bmatrix}.$$

With this definition of λ, (6.2) becomes $\mathbf{A}\mathbf{a} = \lambda\mathbf{a}$. Since $p_A(\lambda) = (2 - \lambda)(1 - \lambda) - 1$, the two eigenvalues are $\lambda_1 \simeq 2.618$ and $\lambda_2 \simeq 0.382$ and correspond to the frequencies of oscillation $\omega_i = \sqrt{K\lambda_i/m}$ which are admitted by our system.   •

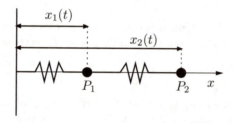

Fig. 6.1. The system of two pointwise bodies of equal mass connected by springs

Problem 6.2 (Demography) Several mathematical models have been proposed in order to predict the evolution of certain species (either human or animal). The simplest population model, which was introduced in 1920 by Lotka and formalized by Leslie 20 years later, is based on the rate of mortality and fecundity for different age intervals, say $i = 0, \ldots, n$. Let $x_i^{(t)}$ denote the number of females (males don't matter in this context) whose age at time t falls in the i-th interval. The values of $x_i^{(0)}$ are given. Moreover, let s_i denote the rate of survival of the females belonging to the i-th interval, and m_i the average number of females generated from a female in the i-th interval.

The model by Lotka and Leslie is described by the set of equations

$$x_{i+1}^{(t+1)} = x_i^{(t)} s_i \qquad i = 0, \ldots, n - 1,$$
$$x_0^{(t+1)} = \sum_{i=0}^{n} x_i^{(t)} m_i.$$

The first n equations describe the population development, the last its reproduction. In matrix form we have

$$\mathbf{x}^{(t+1)} = A\mathbf{x}^{(t)},$$

where $\mathbf{x}^{(t)} = (x_0^{(t)}, \ldots, x_n^{(t)})^T$ and A is the *Leslie matrix*:

$$A = \begin{bmatrix} m_0 & m_1 & \cdots & \cdots & m_n \\ s_0 & 0 & \cdots & \cdots & 0 \\ 0 & s_1 & \ddots & & \vdots \\ \vdots & & \ddots & \ddots & \vdots \\ 0 & 0 & 0 & s_{n-1} & 0 \end{bmatrix}.$$

We will see in Section 6.1 that the dynamics of this population is determined by the eigenvalue λ_1 of A of maximum modulus, whereas the distribution of the individuals in the different age intervals (normalized with respect to the whole population), is obtained as the limit of $\mathbf{x}^{(t)}$ for $t \to \infty$ and satisfies $A\mathbf{x} = \lambda_1 \mathbf{x}$. This problem will be solved in Exercise 6.2.   •

Problem 6.3 (Interurban viability) Let us consider n cities and let A be a matrix whose entry a_{ij} is equal to 1 if the i-th city is connected directly to the j-th city, and 0 otherwise. One can show that the components of the eigenvector $\mathbf{x}$ (of unit length) associated with the maximum eigenvalue provides the accessibility rate (which is a measure of the ease of access) to the various cities. Consider for instance the railway network connecting the 11 main cities of Lombardy in Northern Italy, see Figure 6.2. The moduli of the components of $\mathbf{x}$ in this case are

0.5271	(Milan)	0.1590	(Pavia)
0.2165	(Lodi)	0.3579	(Brescia)
0.4690	(Bergamo)	0.3861	(Como)
0.1590	(Varese)	0.2837	(Lecco)
0.0856	(Sondrio)	0.1906	(Cremona)
0.0575	(Mantua)		

We can see that the cities better connected are (in decreasing order) Milan, Bergamo, Como and Brescia, while Mantua is the worst connected. Note that in spite of the fact that Pavia, Varese, Mantua and Sondrio have the same number of accesses (one), their accessibility rate is different. Obviously, this analysis does not take account at all of the frequency of the connections but merely the existence of the link between the different cities.   •

In the special case where A is either diagonal or triangular, its eigenvalues are nothing but its diagonal entries. However, if A is a general matrix and its dimension n is sufficiently large, seeking the zeros of $p_A(\lambda)$ is not a convenient approach. *Ad hoc* algorithms are better suited, and one of them is described in the next section.

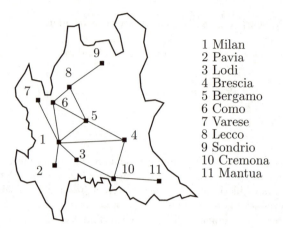

Fig. 6.2. A schematic representation of the railway network between the main cities of Lombardy

6.1 The power method

As noticed in Problems 6.2 and 6.3, the knowledge of the whole *spectrum* of A (that is the the set of all its eigenvalues) is not always required. Often, only the *extremal* eigenvalues matter, that is, those having largest and smallest modulus.

For instance, suppose that A is a square matrix of dimension n, with real entries, and assume that its eigenvalues can be ordered as

$$|\lambda_1| > |\lambda_2| \geq |\lambda_3| \geq \ldots \geq |\lambda_n|. \tag{6.3}$$

Then, in particular, λ_1 is distinct from the other eigenvalues of A. Let us indicate by $\mathbf{x}_1$ the eigenvector (with unit length) associated with λ_1. If the eigenvectors of A are linearly independent, λ_1 and $\mathbf{x}_1$ can be computed by the following iterative procedure, commonly known as the *power method*:

given an arbitrary initial vector $\mathbf{x}^{(0)}$ and setting $\mathbf{y}^{(0)} = \mathbf{x}^{(0)}/\|\mathbf{x}^{(0)}\|$, compute for $k \geq 1$

$$\mathbf{x}^{(k)} = A\mathbf{y}^{(k-1)}, \quad \mathbf{y}^{(k)} = \frac{\mathbf{x}^{(k)}}{\|\mathbf{x}^{(k)}\|}, \quad \lambda^{(k)} = (\mathbf{y}^{(k)})^T A\mathbf{y}^{(k)} \tag{6.4}$$

Note that, by recursion, one finds $\mathbf{y}^{(k)} = \beta^{(k)} A^k \mathbf{y}^{(0)}$ where $\beta^{(k)} = (\Pi_{i=1}^k \|\mathbf{x}^{(i)}\|)^{-1}$ for $k \geq 1$. The presence of the powers of A justifies the name given to this method.

As we will see in next section, this method generates a sequence of vectors $\{\mathbf{y}^{(k)}\}$ with unit length which, as $k \to \infty$, align themselves along the direction of the eigenvector $\mathbf{x}_1$. The error $\|\mathbf{y}^{(k)} - \mathbf{x}_1\|$ is proportional to the ratio $|\lambda_2/\lambda_1|$ in the case of a generic matrix, and to $|\lambda_2/\lambda_1|^2$ when the matrix A is symmetric. Consequently one obtains that $\lambda^{(k)} \to \lambda_1$ for $k \to \infty$.

An implementation of the power method is given in the Program 9. The iterative procedure is stopped at the first iteration k when

$$|\lambda^{(k)} - \lambda^{(k-1)}| < \varepsilon|\lambda^{(k)}|,$$

where ε is a desired tolerance. The input parameters are the matrix A, the initial vector x0, the tolerance tol for the stopping test and the maximum admissible number of iterations nmax. Output parameters are the maximum modulus eigenvalue lambda, the associated eigenvector and the actual number of iterations which have been carried out.

Program 9 - eigpower: power method

```
function [lambda,x,iter]=eigpower(A,tol,nmax,x0)
%EIGPOWER    Numerically evaluate one eigenvalue of a matrix.
%   LAMBDA = EIGPOWER(A) compute with the power method the
%   eigenvalue of A of maximum modulus from an initial guess which by default
%   is an all one vector.
%   LAMBDA = EIGPOWER(A,TOL,NMAX,X0) uses an absolute error
%   tolerance TOL (the default is 1.e-6) and a maximum number of iterations
%   NMAX (the default is 100), starting from the initial vector X0.
%   [LAMBDA,V,ITER] = EIGPOWER(A,TOL,NMAX,X0) also returns the eigenvector
%   V such that A*V=LAMBDA*V and the iteration number at which V was computed.
[n,m] = size(A);
if n ~= m, error('Only for square matrices'); end
if nargin == 1
   tol = 1.e-06;
   x0 = ones(n,1);
   nmax = 100;
end
x0 = x0/norm(x0);
pro = A*x0;
lambda = x0'*pro;
err = tol*abs(lambda) + 1;
iter = 0
while err > tol*abs(lambda) & abs(lambda) ~= 0 & iter <= nmax
   x = pro; x = x/norm(x);
   pro = A*x; lambdanew = x'*pro;
   err = abs(lambdanew - lambda);
   lambda = lambdanew;
```

```
iter = iter + 1;
end
```

Example 6.1 Consider the family of matrices

$$A(\alpha) = \begin{bmatrix} \alpha & 2 & 3 & 13 \\ 5 & 11 & 10 & 8 \\ 9 & 7 & 6 & 12 \\ 4 & 14 & 15 & 1 \end{bmatrix}, \qquad \alpha \in \mathbb{R}.$$

We want to approximate the eigenvalue with largest modulus by the power method. When $\alpha = 30$, the eigenvalues of the matrix are given by $\lambda_1 = 39.396$, $\lambda_2 = 17.8208$, $\lambda_3 = -9.5022$ and $\lambda_4 = 0.2854$ (only the first four significant digits are reported). The method approximates λ_1 in 22 iterations with a tolerance $\varepsilon = 10^{-10}$ and $\mathbf{x}^{(0)} = \mathbf{1}^T$. However, if $\alpha = -30$ we need as many as 708 iterations.

The different behavior can be explained by noting that in the latter case one has $\lambda_1 = -30.643$, $\lambda_2 = 29.7359$, $\lambda_3 = -11.6806$ and $\lambda_4 = 0.5878$. Thus, $|\lambda_2|/|\lambda_1| = 0.9704$, close to unity.

6.1.1   Convergence analysis

Since we have assumed that the eigenvectors $\mathbf{x}_1, \ldots, \mathbf{x}_n$ of A are linearly independent, these eigenvectors form a basis for $\mathbb{C}^n$. Thus the vectors $\mathbf{x}^{(0)}$ and $\mathbf{y}^{(0)}$ can be written as

$$\mathbf{x}^{(0)} = \sum_{i=1}^{n} \alpha_i \mathbf{x}_i, \quad \mathbf{y}^{(0)} = \beta^{(0)} \sum_{i=1}^{n} \alpha_i \mathbf{x}_i, \quad \text{with } \beta^{(0)} = 1/\|\mathbf{x}^{(0)}\|.$$

At the first step the power method gives

$$\mathbf{x}^{(1)} = A\mathbf{y}^{(0)} = \beta^{(0)} A \sum_{i=1}^{n} \alpha_i \mathbf{x}_i = \beta^{(0)} \sum_{i=1}^{n} \alpha_i \lambda_i \mathbf{x}_i$$

and, similarly,

$$\mathbf{y}^{(1)} = \beta^{(1)} \sum_{i=1}^{n} \alpha_i \lambda_i \mathbf{x}_i, \quad \beta^{(1)} = \frac{1}{\|\mathbf{x}^{(0)}\| \, \|\mathbf{x}^{(1)}\|}.$$

At the generic k-th step we will have

$$\mathbf{y}^{(k)} = \beta^{(k)} \sum_{i=1}^{n} \alpha_i \lambda_i^k \mathbf{x}_i, \quad \beta^{(k)} = \frac{1}{\|\mathbf{x}^{(0)}\| \cdots \|\mathbf{x}^{(k)}\|}$$

and therefore

$$\mathbf{y}^{(k)} = \lambda_1^k \beta^{(k)} \left(\alpha_1 \mathbf{x}_1 + \sum_{i=2}^{n} \alpha_i \frac{\lambda_i^k}{\lambda_1^k} \mathbf{x}_i \right).$$

Since $|\lambda_i/\lambda_1| < 1 \; \forall i = 2, \ldots, n$, the vector $\mathbf{y}^{(k)}$ tends to align along the same direction as the eigenvector $\mathbf{x}_1$ when k tends to $+\infty$, provided $\alpha_1 \neq 0$. The condition on α_1, which is impossible to ensure in practice since $\mathbf{x}_1$ is unknown, is in fact not restrictive. Actually, the effect of roundoff errors is the appearance of a non-null component along the direction of $\mathbf{x}_1$, even though this was not the case for the initial vector $\mathbf{x}^{(0)}$. (We can say that this is one of the rare circumstances where roundoff errors help us!)

Example 6.2 Consider the matrix $A(\alpha)$ of Example 6.1, with $\alpha = 16$. The eigenvector $\mathbf{x}_1$ of unit length associated with λ_1 is $(1/2, 1/2, 1/2, 1/2)^T$. Let us choose (on purpose!) the initial vector $(2, -2, 3, -3)^T$, which is orthogonal to $\mathbf{x}_1$. We report in Figure 6.3 the cosine of the angle contained between $\mathbf{y}^{(k)}$ and $\mathbf{x}_1$. We can see that after about 30 iterations of the power method the cosine tends to -1 and the angle tends to π, while the sequence $\lambda^{(k)}$ approaches $\lambda_1 = 34$. The power method has therefore generated, thanks to the roundoff errors, a sequence of vectors $\mathbf{y}^{(k)}$ whose component along the direction of $\mathbf{x}_1$ is increasingly relevant.

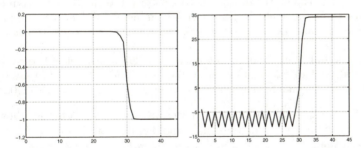

Fig. 6.3. The cosine of the angle contained between $\mathbf{y}^{(k)}$ and $\mathbf{x}_1$ (left), and the value of $\lambda^{(k)}$ (right), for $k = 1, \ldots, 44$

Remark 6.1 It is possible to prove that the power method converges even if λ_1 is a multiple root of $p_A(\lambda)$. On the contrary it does not converge when there exist two distinct eigenvalues both with maximum modulus. In that case the sequence $\lambda^{(k)}$ does not converge to any limit, rather it oscillates between two values.

See the Exercises 6.1-6.3.

6.2   Generalization of the power method

A first possible generalization of the power method consists of applying it to the inverse of the matrix A (provided A is non singular!). Since the eigenvalues of A^{-1} are the reciprocals of those of A, the power method in that case allows us to approximate the eigenvalue of A of minimum modulus. In this way we obtain the so-called *inverse power method*:

given an initial vector $\mathbf{x}^{(0)}$, we set $\mathbf{y}^{(0)} = \mathbf{x}^{(0)}/\|\mathbf{x}^{(0)}\|$ and, for $k \geq 1$, we compute

$$\mathbf{x}^{(k)} = A^{-1}\mathbf{y}^{(k-1)}, \quad \mathbf{y}^{(k)} = \frac{\mathbf{x}^{(k)}}{\|\mathbf{x}^{(k)}\|}, \quad \mu^{(k)} = (\mathbf{y}^{(k)})^T A^{-1}\mathbf{y}^{(k)}$$

If A admits linearly independent eigenvectors, and if also the eigenvalue λ_n of minimum modulus is distinct from the others, then

$$\lim_{k \to \infty} \mu^{(k)} = 1/\lambda_n$$

(*i.e.* $(\mu^{(k)})^{-1}$ tends to λ_n for $k \to \infty$).

At each step k we have to solve a linear system of the form $A\mathbf{x}^{(k)} = \mathbf{y}^{(k-1)}$. It is therefore convenient to generate the LU factorization of A (or its Cholesky factorization if A is symmetric and positive definite) once for all, and then solve two triangular systems at each iteration.

Example 6.3 When applied to the matrix A(30) of Example 6.1, the inverse power method after 7 iterations yields the value 3.5037. Thus the eigenvalue of A(30) of minimum modulus will be approximately equal to $1/3.5037 \simeq 0.2854$.

A further generalization of the power method stems from the following consideration. Let λ_μ denote the (unknown) eigenvalue of A closest to a given number (real or complex) μ. In order to approximate λ_μ, we can at first approximate the minimum length eigenvalue, say $\lambda_{min}(A_\mu)$, of the shifted matrix $A_\mu = A - \mu I$, and then set $\lambda_\mu = \lambda_{min}(A_\mu) + \mu$. We can therefore apply the inverse power method to A_μ to obtain an approximation of $\lambda_{min}(A_\mu)$. This technique is known as the *power method with shift*, and the number μ is called the *shift*.

In the Program 10 we implement the inverse power method with shift. By simply setting $\mu = 0$ we recover the inverse power method. The first 4 input parameters are the same as in Program 9, while mu is the shift. Output parameters are the eigenvalue λ_μ of A, its associated eigenvector x and the actual number of iterations that have been carried out.

Program 10 - invshift: inverse power method with shift

```
function [lambda,x,iter]=invshift(A,mu,tol,nmax,x0)
%INVSHIFT    Numerically evaluate one eigenvalue of a matrix.
%   LAMBDA = INVSHIFT(A) compute with the inverse power method the
%   eigenvalue of A of minimum modulus.
%   LAMBDA = INVSHIFT(A,MU) compute the eigenvalue
%   of A closest to the given number (real or complex) MU.
%   LAMBDA = INVSHIFT(A,MU,TOL,NMAX,X0) uses an absolute error
%   tolerance TOL (the default is 1.e-6) and a maximum number of iterations
%   NMAX (the default is 100), starting from the initial vector X0.
%   [LAMBDA,V,ITER] = INVSHIFT(A,MU,TOL,NMAX,X0) also returns the eigenvector
%   V such that A*V=LAMBDA*V and the iteration number at which V was computed.
[n,m]=size(A);
if n ~= m, error('Only for square matrices'); end
if nargin == 1
  x0 = rand(n,1); nmax = 100; tol = 1.e-06; mu = 0;
elseif nargin == 2
  x0 = rand(n,1); nmax = 100; tol = 1.e-06;
end
[L,U]=lu(A-mu*eye(n));
if norm(x0) == 0
  x0 = rand(n,1);
end
x0=x0/norm(x0);
z0=L\x0;
pro=U\z0;
lambda=x0'*pro;
err=tol*abs(lambda)+1;      iter=0;
while err > tol*abs(lambda) & abs(lambda) ~= 0 & iter <= nmax
   x = pro; x = x/norm(x);
   z=L\x;    pro=U\z;
   lambdanew = x'*pro;
   err = abs(lambdanew - lambda);
   lambda = lambdanew;
   iter = iter + 1;
end
lambda = 1/lambda + mu;
```

Example 6.4 For the matrix $A(30)$ of Example 6.1 we seek the eigenvalue closest to the value 17. For that we use the Program 10 with mu=17, tol $=10^{-10}$ and x0=[1;1;1;1]. After 8 iterations the Program returns the value lambda=1.2183. A less accurate knowledge of the *shift* would involve more iterations. For instance, if we set `mu=13` the program will return the value 17.8208 after 19 iterations.

The value of the shift can be modified during the iterations, by setting $\mu = \lambda^{(k)}$. This yields a faster convergence; however the computational cost grows substantially since now at each iteration the matrix A_μ does change.

See the Exercises 6.4-6.6.

6.3   How to compute the shift

In order to apply successfully the power method with shift we need to locate (more or less accurately) the eigenvalues of A in the complex plane. To this end let us introduce the following definition.

Let A be a square matrix of dimension n. The *Gershgorin circles* $C_i^{(r)}$ and $C_i^{(c)}$ associated with its i-th row and i-th column are respectively defined as

$$C_i^{(r)} = \{z \in \mathbb{C} : |z - a_{ii}| \leq \sum_{j=1, j\neq i}^{n} |a_{ij}|\},$$
$$C_i^{(c)} = \{z \in \mathbb{C} : |z - a_{ii}| \leq \sum_{j=1, j\neq i}^{n} |a_{ji}|\}.$$

$C_i^{(r)}$ is called the i-th *row circle* and $C_i^{(c)}$ the i-th *column circle*.

hold
on/off

By the Program 11 we can visualize in two different windows the row circles and the column circles of a matrix. The command `hold on` allows the overlapping of subsequent pictures (in our case, the different circles that have been computed in sequential mode). This command can be neutralized by the command `hold off`.

patch

axis

The command `patch` was used in order to color the circles, while the command `axis` sets scaling for the x- and y-axes on the current plot.

Program 11 - gershcircles: Gershgorin circles

```
function gershcircles(A)
%GERSHCIRCLES plots the Gershgorin circles
%   GERSHGORINCIRCLES(A) plots the Gershgorin circles for
%   the square matrix A and its transpose.
Abs = abs(A); n = max(size(A)); radii = sum(Abs,2)-diag(Abs);
xcenter = real(diag(A)); ycenter = imag(diag(A)); theta = [0:pi/100:2*pi];
```

```
costheta = cos(theta); sintheta = sin(theta);
x=[]; y=[]; figure(1); clf; axis equal;  hold on;
for i = 1:n
 x=[x; radii(i)*cos(theta)+xcenter(i)];y=[y; radii(i)*sin(theta)+ycenter(i)];
 patch(x(i,:),y(i,:),'red');
end
for i = 1:n, plot(x(i,:),y(i,:),'k',xcenter(i),ycenter(i),'xk'), end
xmax = max(max(x)); ymax=max(max(y));
xmin = min(min(x)); ymin=min(min(y));
hold off; figure(2); clf; axis equal; hold on; radii = sum(Abs)-(diag(Abs))';
x=[]; y=[]; clf; axis equal; hold on;
for i = 1:n
 x=[x; radii(i)*cos(theta)+xcenter(i)];y=[y; radii(i)*sin(theta)+ycenter(i)];
 patch(x(i,:),y(i,:),'green')
end
for i = 1:n, plot(x(i,:),y(i,:),'k',xcenter(i),ycenter(i),'xk'); end
xmax = max(max(max(x)),xmax); ymax=max(max(max(y)),ymax);
xmin = min(min(min(x)),xmin); ymin=min(min(min(y)),ymin); hold off;
axis([xmin xmax ymin ymax]); figure(1); axis([xmin xmax ymin ymax]);
```

Example 6.5 In Figure 6.4 we have plotted the Gershgorin circles associated with the matrix

$$A = \begin{bmatrix} 30 & 1 & 2 & 3 \\ 4 & 15 & -4 & -2 \\ -1 & 0 & 3 & 5 \\ -3 & 5 & 0 & -1 \end{bmatrix}.$$

The centers of the circles have been identified by dots.

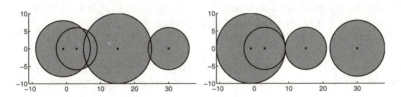

Fig. 6.4. Row circles (left) and column circles (right) for the matrix of Example 6.5

As previously anticipated, Gershgorin circles may be used to locate the eigenvalues of a matrix, as stated in the following proposition.

Proposition 6.1 *All eigenvalues of a given matrix* $A \in \mathbb{C}^{n \times n}$ *belong to the region of the complex plane which is the intersection of the two* *regions formed respectively by the union of the row circles and the union of the column circles.*

Moreover, should m row circles (or column circles), with $1 \le m \le n$, be disconnected from the union of the remaining $n - m$ circles, then their union contains exactly m eigenvalues.

There is no guarantee that a circle should contain eigenvalues, unless it is isolated from the others. The previous result can be applied in order to obtain a preliminary guess of the shift, as we show in the following example.

Example 6.6 From the analysis of the row circles of the matrix A(30) of Example 6.1 we deduce that the real parts of the eigenvalues of A lie between -32 and 48. Thus we can use Program 10 to compute the maximum modulus eigenvalue by setting the value of the shift μ equal to 48. The convergence is achieved in 16 iterations, whereas 24 iterations would be required using the power method with the same initial guess x0=[1;1;1;1] and the same tolerance tol=1.e-10.

See the Exercises 6.7-6.8.

6.4   Computation of all the eigenvalues

Two square matrices A and B having the same dimension are called *similar* if there exists a non singular matrix P such that

$$P^{-1}AP = B.$$

Similar matrices share the same eigenvalues. Indeed, if λ is an eigenvalue of A and $\mathbf{x} \ne \mathbf{0}$ is an associated eigenvector, we have

$$B P^{-1}\mathbf{x} = P^{-1}A\mathbf{x} = \lambda P^{-1}\mathbf{x},$$

that is, λ is also an eigenvalue of B and its associated eigenvector is now $\mathbf{y} = P^{-1}\mathbf{x}$.

The methods which allow a simultaneous approximation of all the eigenvalues of a matrix are generally based on the idea of transforming A (after an infinite number of steps) into a similar matrix with diagonal or triangular form, whose eigenvalues are therefore given by the entries lying on its main diagonal.

Among these methods we mention the *QR method* which is imple-
eig   mented in MATLAB in the function eig. More precisely, the command
D=eig(A) returns a vector D containing all the eigenvalues of A. However,

by setting [X,D]=eig(A), we obtain two matrices: the diagonal matrix
D formed by the eigenvalues of A, and a matrix X whose column vectors
are the eigenvectors of A. Thus, A*X=X*D. When A is stored in a sparse
mode the command eigs(A,k) allows the computation of the first k eigs
eigenvalues of larger modulus of A.

Should all eigenvalues of a matrix A be distinct, the sequence of eigenvalues generated by the QR method converges to an upper triangular
matrix similar to A, which reduces to a diagonal matrix when A is symmetric. A description of the algorithm which stands at the basis of the
QR method is provided in the Exercises 6.9 and 6.10 and, more thor-
oughly, in [QSS00], Chapter 5.

Let us summarize

1. The power method is an iterative procedure to compute the eigenvalue of maximum modulus of a given matrix;

2. the inverse power method allows the computation of the eigenvalue
   of minimum modulus; it requires the factorization of the given
   matrix;

3. the power method with shift allows the computation of the eigenvalue closest to a given number; its effective application requires
   some *a-priori* knowledge of the location of the eigenvalues of the
   matrix, which can be achieved inspecting the Gershgorin circles;

4. the QR method is a global technique which allows the simultaneous
   approximation of all the eigenvalues of a given matrix.

See the Exercises 6.9-6.11.

6.5 What we haven't told you

We have not analyzed the issue of the condition number of the eigenvalue problem, which measures the sensitivity of the eigenvalues to the
variation of the entries of the matrix. The interested reader is referred
to, for instance, [Wil65], [GL89] and [QSS00], Chapter 5.

Let us just remark that the eigenvalue computation is not necessarily
an ill conditioned problem when the condition number of the matrix is
large. An instance of this is provided by the Hilbert matrix (see Example
5.8): although its condition number is extremely large, the eigenvalue

computation of the Hilbert matrix is well conditioned thanks to the fact that the matrix is symmetric and positive definite.

Besides the QR method, for computing simultaneously all the eigenvalues, we can use the Jacobi method which transforms a symmetric matrix into a diagonal matrix, by eliminating, step-by-step, through similarity transformations, every off-diagonal element. This method does not terminate in a finite number of steps since, while a new off-diagonal element is set to zero, those previously treated can reassume non-zero values.

Other methods are the Lanczos method and the method which uses the so-called Sturm sequences. For a survey of all these methods see [Saa92].

The MATLAB library ARPACKC (available through the command arpackc arpackc) can be used to compute the eigenvalues of large matrices, and can be download from the address

$$\text{http://www.caam.rice.edu/software/ARPACK/}$$

eigs is a command that uses this library.

Let us mention that an appropriate use of the *deflation* technique (which consists of a successive elimination of the eigenvalues already computed) allows the acceleration of the convergence of the previous methods and hence the reduction of their computational cost.

6.6   Exercises

Exercise 6.1 Upon setting the tolerance equal to $\varepsilon = 10^{-10}$, use the power method to approximate the maximum modulus eigenvalue for the following matrices, starting from the initial vector $\mathbf{x}^{(0)} = (1, 2, 3)^T$:

$$A_1 = \begin{bmatrix} 1 & 2 & 0 \\ 1 & 0 & 0 \\ 0 & 1 & 0 \end{bmatrix}, \quad A_2 = \begin{bmatrix} 0.1 & 3.8 & 0 \\ 1 & 0 & 0 \\ 0 & 1 & 0 \end{bmatrix}, \quad A_3 = \begin{bmatrix} 0 & -1 & 0 \\ 1 & 0 & 0 \\ 0 & 1 & 0 \end{bmatrix}.$$

Then comment on the convergence behavior of the method in the three different cases.

Exercise 6.2 The features of a population of fishes are described by the following Leslie matrix introduced in Problem 6.2:

Age interval (months)	$\mathbf{x}^{(0)}$	m_i	s_i
0-3	6	0	0.2
3-6	12	0.5	0.4
6-9	8	0.8	0.8
9-12	4	0.3	–

Find the vector $\mathbf{x}$ of the normalized distribution of this population for different age intervals, according to what we have seen in Problem 6.2.

Exercise 6.3 Prove that the power method does not converge for matrices featuring an eigenvalue of maximum modulus $\lambda_1 = \gamma e^{i\vartheta}$ and another eigenvalue $\lambda_2 = \gamma e^{-i\vartheta}$, where $i = \sqrt{-1}$ and $\gamma, \vartheta \in \mathbb{R}$.

Exercise 6.4 Show that the eigenvalues of A^{-1} are the reciprocals of those of A.

Exercise 6.5 Verify that the power method is unable to compute the maximum modulus eigenvalue of the following matrix, and explain why:

$$A = \begin{bmatrix} \frac{1}{3} & \frac{2}{3} & 2 & 3 \\ 1 & 0 & -1 & 2 \\ 0 & 0 & -\frac{5}{3} & -\frac{2}{3} \\ 0 & 0 & 1 & 0 \end{bmatrix}.$$

Exercise 6.6 By using the power method with shift, compute the largest positive eigenvalue and the largest negative eigenvalue of

$$A = \begin{bmatrix} 3 & 1 & 0 & 0 & 0 & 0 & 0 \\ 1 & 2 & 1 & 0 & 0 & 0 & 0 \\ 0 & 1 & 1 & 1 & 0 & 0 & 0 \\ 0 & 0 & 1 & 0 & 1 & 0 & 0 \\ 0 & 0 & 0 & 1 & 1 & 1 & 0 \\ 0 & 0 & 0 & 0 & 1 & 2 & 1 \\ 0 & 0 & 0 & 0 & 0 & 1 & 3 \end{bmatrix}.$$

A is the so-called *Wilkinson matrix* and can be generated by the command `wilkinson(7)`.   `wilkinson`

Exercise 6.7 By using the Gershgorin circles, provide an estimate of the maximum number of the complex eigenvalues of the following matrices:

$$A = \begin{bmatrix} 2 & -\frac{1}{2} & 0 & -\frac{1}{2} \\ 0 & 4 & 0 & 2 \\ -\frac{1}{2} & 0 & 6 & \frac{1}{2} \\ 0 & 0 & 1 & 9 \end{bmatrix}, B = \begin{bmatrix} -5 & 0 & 1/2 & 1/2 \\ 1/2 & 2 & 1/2 & 0 \\ 0 & 1 & 0 & 1/2 \\ 0 & 1/4 & 1/2 & 3 \end{bmatrix}.$$

Exercise 6.8 Use the result of Proposition 6.1 to find a suitable shift for the computation of the maximum modulus eigenvalue of

$$A = \begin{bmatrix} 5 & 0 & 1 & -1 \\ 0 & 2 & 0 & -\frac{1}{2} \\ 0 & 1 & -1 & 1 \\ -1 & -1 & 0 & 0 \end{bmatrix}.$$

Then compare the number of iterations as well the computational cost of the power method both with and without shift by setting the tolerance equal to 10^{-14}.

Exercise 6.9 A matrix $A \in \mathbb{R}^{n \times n}$ admits a QR factorization if there exist a matrix $Q \in \mathbb{R}^{n \times n}$, with the orthogonality property $Q^T Q = I$, and an upper triangular matrix $R \in \mathbb{R}^{n \times n}$, such that $A = QR$. This factorization can be qr   obtained by the command [Q,R]=qr(A). Use this command to obtain the QR factorization of the matrix A of Exercise 6.7. Then verify that the matrix $C = RQ$ is similar to A.

Exercise 6.10 The following algorithm provides the basis for the so-called *QR method* for computing all the eigenvalues of a matrix A:
set $A^{(0)} = A$;
    then for $k = 0, 1, \ldots$, compute $Q^{(k)}$ and $R^{(k)}$ such that $A^{(k)} = Q^{(k)}R^{(k)}$;
next define $A^{(k+1)} = R^{(k)}Q^{(k)}$.
    Write a MATLAB program that implements this method. Then carry out a few iterations on the matrix A of Exercise 6.7. What is the limit of the sequence of matrices $A^{(k)}$?

Exercise 6.11 Use the command **eig** to compute all the eigenvalues of the two matrices given in Exercise 6.7. Then check how accurate are the conclusions drawn on the basis of Proposition 6.1.

7. Ordinary differential equations

A differential equation is an equation involving one or more derivatives of an unknown function. If all derivatives are taken with respect to a single independent variable we call it an *ordinary differential equation*, whereas we have a *partial differential equation* when partial derivatives are present.

The differential equation (ordinary or partial) has *order* p if p the maximum order of differentiation that is present. The next chapter will be devoted to the study of a particular kind of second order partial differential equations, the elliptic equations, whereas in the present chapter we will deal with ordinary differential equations of first order.

Ordinary differential equations can describe the evolution of many phenomena in various fields, as we can see from the following three examples.

Problem 7.1 (Thermodynamics) Consider a body having internal temperature T which is set in an environment with constant temperature T_e. Assume that its mass m is concentrated in a single point. Then the heat transfer between the body and the external environment can be described by the Stefan-Boltzmann law

$$v(t) = \epsilon \gamma S(T^4(t) - T_e^4)$$

where t is the time variable, ϵ the Boltzmann constant (equal to $5.6 \cdot 10^{-8} \, \mathrm{J/m^2 K^4 s}$ where J stands for Joule, K for Kelvin and, obviously, m for meter, s for second), γ is the emissivity constant of the body, S the area of its surface and v is the rate of the heat transfer. The rate of variation of the energy $E(t) = mCT(t)$ (where C denotes the specific heat of the material constituting the body) equals, in absolute value, the rate v. Consequently, setting $T(0) = T_0$, the computation of $T(t)$ requires the solution of the ordinary differential equation

$$\frac{dT}{dt} = -\frac{v(t)}{mC}. \tag{7.1}$$

See Exercise 7.16.   •

Problem 7.2 (Biology) Consider a population of bacteria in a confined environment in which no more than B elements can coexist. Assume that, at the initial time, the number of individuals is equal to $y_0 \ll B$ and the growth rate of the bacteria is a positive constant C. In this case the rate of change of the population is proportional to the number of existing bacteria, under the restriction that the total number cannot exceed B. This is expressed by the differential equation

$$\frac{dy}{dt} = Cy \left(1 - \frac{y}{B}\right),\tag{7.2}$$

whose solution $y = y(t)$ denotes the number of bacteria at time t.

Assuming that two populations y_1 and y_2 be in competition, instead of (7.2) we would have

$$\frac{dy_1}{dt} = C_1 y_1 \left(1 - b_1 y_1 - d_2 y_2\right),$$
$$\frac{dy_2}{dt} = -C_2 y_2 \left(1 - b_2 y_2 - d_1 y_1\right),\tag{7.3}$$

where C_1 and C_2 represent the growth rates of the two populations. The coefficients d_1 and d_2 govern the type of interaction between the two populations, while b_1 and b_2 are related to the available quantity of nutrients. The above equations (7.3) are called the Lotka-Volterra equations and form the basis of various applications. For their numerical solution, see Example 7.6.   •

Problem 7.3 (Electricity) Consider the electrical circuit of Figure 7.1. We want to compute the function $v(t)$ representing the potential drop at the ends of the capacitor C starting from the initial time $t = 0$ at which the switch I has been turned off. Assume that the inductance L can be expressed as an explicit function of the current intensity i, that is $L = L(i)$. The Ohm law yields

$$e - \frac{d(i_1 L(i_1))}{dt} = i_1 R_1 + v,$$

where R_1 is a resistance. By assuming the current fluxes to be directed as indicated in Figure 7.1, upon differentiating with respect to t both sides of the Kirchoff law $i_1 = i_2 + i_3$ and noticing that $i_3 = Cdv/dt$ and $i_2 = v/R_2$, we find the further equation

$$\frac{di_1}{dt} = C\frac{d^2v}{dt^2} + \frac{1}{R_2}\frac{dv}{dt}.$$

We have therefore found a system of two differential equations whose solution allows the description of the time variation of the two unknowns i_1 and v. The second equation has order 2. For its solution see Example 7.7.   •

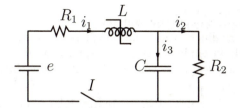

Fig. 7.1. The electrical circuit of Problem 7.3

7.1   The Cauchy problem

We confine ourselves to first order differential equations, as an equation of order $p > 1$ can always be reduced to a system of p equations of order 1. The case of first order systems will be addressed in Section 7.8.

An ordinary differential equation in general admits an infinity of solutions. In order to fix one of them we must impose a further condition which prescribes the value taken by this solution at a given point of the integration interval. For instance, the equation (7.2) admits the family of solutions $y(t) = B\psi(t)/(1 + \psi(t))$ with $\psi(t) = e^{Ct+K}$, K being an arbitrary constant. If we impose the condition $y(0) = 1$, we pick up the unique solution corresponding to the value $K = \ln[1/(B-1)]$.

We will therefore consider the solution of the so-called *Cauchy problem* which takes the following form:

find $y : I \to \mathbb{R}$ such that

$$\begin{cases} y'(t) = f(t, y(t)) & \forall t \in I, \\ y(t_0) = y_0, \end{cases} \tag{7.4}$$

where I is an interval of $\mathbb{R}$, $f : I \times \mathbb{R} \to \mathbb{R}$ is a given function and y' denotes the derivative of y with respect to t. Finally, t_0 is a point of I and y_0 a given value which is called the *initial data*.

In the following proposition we report a classical result of Analysis.

Proposition 7.1 *Assume that the function $f(t,y)$ is*

1. *continuous with respect to both arguments;*

2. *Lipschitz-continuous with respect to its second argument, that is, there exists a positive constant L such that*

$$|f(t, y_1) - f(t, y_2)| \leq L|y_1 - y_2|, \quad \forall t \in I, \ \forall y_1, y_2 \in \mathbb{R}.$$

Then the solution of the Cauchy problem (7.4) exists and is unique.

Unfortunately, explicit solutions are available only for very special types of ordinary differential equations. In some other cases, the solution is available only in implicit form. This is, for instance, the case with the equation $y' = (y - t)/(y + t)$ whose solution satisfies the implicit relation

$$\frac{1}{2}\ln(t^2 + y^2) + \text{arctg}\frac{y}{t} = C,$$

where C is an arbitrary constant. In some other circumstances the solution is not even representable in implicit form, as in the case of the equation $y' = e^{-t^2}$ whose general solution can only be expressed through a series expansion.

For all these reasons, we seek numerical methods capable of approximating the solution of *every* family of ordinary differential equations for which solutions do exist.

The common strategy of all these methods consists of subdividing the integration interval $I = [t_0, T]$, with $T < +\infty$, into N_h intervals of length $h = (T - t_0)/N_h$; h is called the *discretization step*. Then, at each *node* t_n ($0 \leq n \leq N_h - 1$) we seek the unknown value u_n which approximates $y_n = y(t_n)$. The set of values $\{u_0 = y_0, u_1, \ldots, u_{N_h}\}$ is our *numerical solution*.

7.2   Euler methods

A classical method, the *forward Euler* method, generates the numerical solution as follows

$$u_{n+1} = u_n + hf_n, \quad n = 0, \ldots, N_h - 1 \qquad (7.5)$$

where we have used the shorthand notation $f_n = f(t_n, u_n)$. This method is obtained by considering the differential equation (7.4) at every node t_n, $n = 1, \ldots, N_h$ and replacing the exact derivative $y'(t_n)$ by means of the incremental ratio (4.3).

In a similar way, using this time the incremental ratio (4.7) to approximate $y'(t_{n+1})$, we obtain the *backward Euler* method

$$u_{n+1} = u_n + hf_{n+1}, \quad n = 0, \ldots, N_h - 1 \qquad (7.6)$$

Both methods provide an instance of a *one-step method* since for computing the numerical solution u_{n+1} at the node t_{n+1} we only need the information related to the previous node t_n.

More precisely, in the forward Euler method u_{n+1} depends exclusively on the value u_n previously computed, whereas in the backward Euler method it depends also on itself through the value f_{n+1}. For this reason the first method is called the *explicit* Euler method and the second the *implicit* Euler method.

For instance, the discretization of (7.2) by the forward Euler method requires at every step the simple computation of

$$u_{n+1} = u_n + hCu_n\left(1 - u_n/B\right),$$

whereas using the backward Euler method we must solve the non-linear equation

$$u_{n+1} = u_n + hCu_{n+1}\left(1 - u_{n+1}/B\right).$$

Thus, implicit methods are more costly than explicit methods, since at every time-level t_{n+1} we must solve a non-linear problem to compute u_{n+1}. However, we will see that implicit methods enjoy better stability properties than explicit ones.

The forward Euler method is implemented in the Program 12; the integration interval is `tspan = [t0,tfinal]`, `odefun` is a string which contains the function $f(t, y(t))$ which depends on the variables `t` and `y`, or an inline function whose first two arguments stand for t and y.

Program 12 - feuler: forward Euler method

```
function [t,y]=feuler(odefun,tspan,y,Nh,varargin)
%FEULER  Solve differential equations using the forward Euler method.
%   [T,Y] = FEULER(ODEFUN,TSPAN,Y0,NH) with TSPAN = [T0 TFINAL]
%   integrates the system of differential equations y' = f(t,y) from
%   time T0 to TFINAL with initial conditions Y0 using the forward Euler
%   method on an equispaced grid of NH intervals.
%   Function ODEFUN(T,Y) must return a column vector
%   corresponding to f(t,y). Each row in the solution array Y corresponds to
%   a time returned in the column vector T.
%   [T,Y] = FEULER(ODEFUN,TSPAN,Y0,NH,P1,P2,...) passes the additional
%   parameters P1,P2,... to the function ODEFUN as ODEFUN(T,Y,P1,P2...).
h=(tspan(2)-tspan(1))/Nh;
tt=linspace(tspan(1),tspan(2),Nh+1);
for t = tt(1:end-1)
   y = [y; y(end,:) + h*feval(odefun,t,y(end,:),varargin{:})];
end
t=tt;
```

The backward Euler method is implemented in the Program 13. Note that we have used the function `fzero` for the solution of the non-linear problem at each step. As initial data for `fzero` we use the last computed value of the numerical solution. Note the use of the command `num2str(num,s)` which serves the purpose of transforming a number `num` in a string by keeping 16 significant digits.

Program 13 - beuler: backward Euler method

```
function [t,u]=beuler(odefun,tspan,y,Nh,varargin)
%BEULER  Solve differential equations using the backward Euler method.
%   [T,Y] = BEULER(ODEFUN,TSPAN,Y0,NH) with TSPAN = [T0 TFINAL]
%   integrates the system of differential equations y' = f(t,y) from
%   time T0 to TFINAL with initial conditions Y0 using the backward Euler
%   method on an equispaced grid of NH intervals.
%   Function ODEFUN(T,Y) must return a column vector
%   corresponding to f(t,y). Each row in the solution array Y corresponds to
%   a time returned in the column vector T.
%   [T,Y] = BEULER(ODEFUN,TSPAN,Y0,NH,P1,P2,...) passes the additional
%   parameters P1,P2,... to the function ODEFUN as ODEFUN(T,Y,P1,P2...).
h=(tspan(2)-tspan(1))/Nh;
tt=linspace(tspan(1),tspan(2),Nh+1);
u(1)=y;
syms x;
y = x;
for t=tt(2:end)
   fun = inline(['x-',num2str(h,16),'*(',char(feval(odefun,t,y,varargin{:})),...
        ')-',num2str(u(end),16)],'x');
   u = [u, fzero(fun,u(end))];
end
t=tt;
```

7.2.1 Convergence analysis

A numerical method is *convergent* if

$$\forall n = 0, \ldots, N_h, \qquad |u_n - y_n| \le C(h)$$

where $C(h)$ is infinitesimal with respect to h when h tends to zero. If $C(h) = \mathcal{O}(h^p)$ for some $p > 0$, then we say that the method converges with *order* p. In order to verify that the forward Euler method converges, we write the error as follows:

$$e_n = u_n - y_n = (u_n - u_n^*) + (u_n^* - y_n), \qquad (7.7)$$

where $y_n = y(t_n)$ and

$$u_n^* = y_{n-1} + hf(t_{n-1}, y_{n-1}),$$

that is, u_n^* denotes the numerical solution at time t_n which we would obtain starting from the exact solution at time t_{n-1}; see Figure 7.2. The term $u_n^* - y_n$ in (7.7) represents the error produced by a single step of the forward Euler method, whereas the term $u_n - u_n^*$ represents the propagation from t_{n-1} to t_n of the error accumulated at the previous time-level t_{n-1}. The method converges provided both terms tend to zero as $h \to 0$. Assuming that the second order derivative of y exists and is continuous, thanks to (4.5) we find

$$u_n^* - y_n = \frac{h^2}{2} y''(\xi_n), \quad \text{for a suitable } \xi_n \in (t_{n-1}, t_n). \tag{7.8}$$

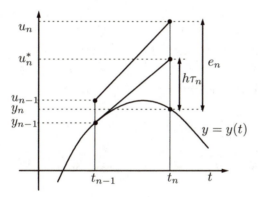

Fig. 7.2. Geometrical representation of a step of the forward Euler method

Consequently, $u_n^* - y_n$ tends to zero as $h \to 0$.

The quantity $\tau_n(h) = (u_n^* - y_n)/h$ represents the error that would be generated by forcing the exact solution to satisfy the numerical scheme, and for this reason is called the *local truncation error*. The *(global) truncation error* is defined as

$$\tau(h) = \max_{n=0,\dots,N_h} |\tau_n(h)|.$$

We note that these definitions of local and global truncation errors hold for a general numerical method, not only for the Euler method.

In view of (7.8), the truncation error for the forward Euler method takes the following form:

$$\tau(h) = Mh/2, \tag{7.9}$$

where $M = \max_{t \in [t_0, T]} |f'(t, y(t))|$.

From (7.8) we deduce that $\lim_{h \to 0} \tau(h) = 0$, and a method for which this happens is said to be *consistent*. Further, we say that it is consistent with order p if $\tau(h) = \mathcal{O}(h^p)$ for a suitable integer $p \geq 1$.

Consider now the other term in (7.7). We have

$$u_n^* - u_n = e_{n-1} + h \left[f(t_{n-1}, y_{n-1}) - f(t_{n-1}, u_{n-1}) \right]. \tag{7.10}$$

Since f is Lipschitz continuous with respect to its second argument, we obtain

$$|u_n^* - u_n| \leq (1 + hL)|e_{n-1}|.$$

If $e_0 = 0$, the previous relations yield

$$
\begin{aligned}
|e_n| &\leq |u_n - u_n^*| + |u_n^* - y_n| \\
&\leq (1 + hL)|e_{n-1}| + h\tau_n(h) \\
&\leq \left[1 + (1 + hL) + \ldots + (1 + hL)^{n-1} \right] h\tau(h) \\
&= \frac{(1 + hL)^n - 1}{L} \tau(h) \leq \frac{e^{L(t_n - t_0)} - 1}{L} \tau(h).
\end{aligned}
$$

We have used the identity

$$\sum_{k=0}^{n-1} (1 + hL)^k = [(1 + hL)^n - 1]/hL,$$

the inequality $1 + hL \leq e^{hL}$ and we have observed that $nh = t_n - t_0$. Therefore we find

$$|e_n| \leq \frac{e^{L(t_n - t_0)} - 1}{L} \frac{M}{2} h, \quad \forall n = 0, \ldots, N_h, \tag{7.11}$$

and thus we can conclude that *the forward Euler method converges with order 1*. We can note that the order of this method coincides with the order of its local truncation error. This property is shared by many numerical methods for the numerical solution of ordinary differential equations.

Remark 7.1 The convergence estimate (7.11) is obtained by simply requiring f to be Lipschitz continuous. A better estimate, precisely $|e_n| \leq Mh(t_n - t_0)/2$, holds if f satisfies the further requirement that $\partial f(t, y)/\partial y \leq 0$ for all $t \in [t_0, T]$ and all $-\infty < y < \infty$. Indeed, in that case, using Taylor expansion, from (7.10) we obtain $u_n^* - u_n = (1 + h\partial f/\partial y(t_{n-1}, \eta_n))e_{n-1}$, thus $|u_n^* - u_n| \leq |e_{n-1}|$, provided the stability inequality $h < 2/\max_t |\partial f/\partial y(t, y(t))|$ holds. Then $|e_n| \leq |u_n^* - u_n| + |e_{n-1}| \leq |e_0| + nh\tau(h)$, whence we conclude owing to (7.9).

Remark 7.2 (Consistency) The property of consistency is necessary in order to get convergence. Actually, should it be violated, at each step the numerical method would generate an error which is not infinitesimal with respect to h. The accumulation with the previous errors would inhibit the global error to converge to zero when $h \to 0$.

Using a similar argument we can also prove that the backward Euler method converges with order 1 with respect to h.

Example 7.1 Consider the Cauchy problem

$$\begin{cases} y'(t) = \cos(2y(t)) & t \in (0,1], \\ y(0) = 0, \end{cases} \qquad (7.12)$$

whose solution is $y(t) = \frac{1}{2}\arcsin((e^{4t} - 1)/(e^{4t} + 1))$. We solve it by the forward Euler method (Program 12) and the backward Euler method (Program 13). By the following commands we use different values of h: $1/2, 1/4, 1/8, \ldots, 1/512$:

```
>> tspan=[0,1]; y0=0; f=inline('cos(2*y)','t','y');
>> u=inline('0.5*asin((exp(4*t)-1)./(exp(4*t)+1))','t');
>> Nh=2;
for k=1:10
    [t,ufe]=feuler(f,tspan,y0,Nh); fe(k)=abs(ufe(end)-feval(u,t(end)));
    [t,ube]=beuler(f,tspan,y0,Nh); be(k)=abs(ube(end)-feval(u,t(end)));
    Nh = 2*Nh;
end
```

The errors committed at the point $t = 1$ are stored in the variable `fe` (forward Euler) and `be` (backward Euler), respectively. Then we apply formula (1.11) to estimate the order of convergence. Using the following commands

```
>> p=log(abs(fe(1:end-1)./fe(2:end)))/log(2); p(1:2:end)
   1.2898    1.0349    1.0080    1.0019    1.0005
>> p=log(abs(be(1:end-1)./be(2:end)))/log(2); p(1:2:end)
   0.9070    0.9720    0.9925    0.9981    0.9995
```

we can verify that both methods are convergent with order 1.

Remark 7.3 The error estimate (7.11) was derived by assuming that the numerical solution $\{u_n\}$ is obtained in exact arithmetic. Should we account for the (inevitable) roundoff-errors, the error might blow up as h approaches 0 as $\mathcal{O}(1/h)$ (see, e.g., [Atk89]). This circumstance suggests that it might be unreasonable to go below a certain threshold h^* (which is actually extremely tiny) in practical computations.

See the Exercises 7.1-7.3.

7.3 The Crank-Nicolson method

Adding together the generic steps of the forward and backward Euler methods we find the so-called *Crank-Nicolson method*

$$u_{n+1} = u_n + \frac{h}{2}[f_n + f_{n+1}]$$

(7.13)

It is a one-step implicit method, which is implemented in the Program 14. Input and output parameters are the same as in the Euler methods.

Program 14 - cranknic: Crank-Nicolson method

```
function [t,u]=cranknic(odefun,tspan,y,Nh,varargin)
%CRANKNIC  Solve differential equations using the Crank-Nicolson method.
%   [T,Y] = CRANKNIC(ODEFUN,TSPAN,Y0,NH) with TSPAN = [T0 TFINAL]
%   integrates the system of differential equations y' = f(t,y) from
%   time T0 to TFINAL with initial conditions Y0 using the Crank-Nicolson
%   method on an equispaced grid of NH intervals.
%   Function ODEFUN(T,Y) must return a column vector
%   corresponding to f(t,y). Each row in the solution array Y corresponds to
%   a time returned in the column vector T.
%   [T,Y] = CRANKNIC(ODEFUN,TSPAN,Y0,NH,P1,P2,...) passes the additional
%   parameters P1,P2,... to the function ODEFUN as ODEFUN(T,Y,P1,P2...).
h=(tspan(2)-tspan(1))/Nh;
tt=linspace(tspan(1),tspan(2),Nh+1);
u(1)=y;
fold = feval(odefun,tspan(1),y,varargin{:});
syms x;
y = x;
for t=tt(2:end)
   y = x;
   fun = inline(['x-',num2str(h,16),'*0.5*(',char(feval(odefun,t,y,varargin{:})),')+(',...
         num2str(fold,16),'))-(',num2str(u(end),16),')'],'x');
   u = [u, fzero(fun,u(end))];
   fold = feval(odefun,t,u(end),varargin{:});
end
t=tt;
```

The local truncation error of the Crank-Nicolson method satisfies

$$h\tau_n(h) = y(t_n) - y(t_{n-1}) - \frac{h}{2}\left[f(t_n, y(t_n)) + f(t_{n-1}, y(t_{n-1}))\right]$$

$$= \int_{t_{n-1}}^{t_n} f(t, y(t))\, dt - \frac{h}{2}\left[f(t_n, y(t_n)) + f(t_{n-1}, y(t_{n-1}))\right].$$

The last equality follows from the fundamental theorem of integration (which we have recalled in Section 1.4.3). The second term expresses the error associated with the trapezoidal rule for numerical integration (4.17). If we assume that $y \in C^3$ and use (4.18), we deduce that

$$\tau_n(h) = -\frac{h^2}{12}y'''(\xi_n) \quad \text{for a suitable } \xi_n \in (t_{n-1}, t_n). \tag{7.14}$$

Thus the Crank-Nicolson method is consistent with order 2, *i.e.* its local truncation error tends to 0 as h^2. Using a similar approach to that followed for the forward Euler method, we can show that the Crank-Nicolson method is convergent with order 2 with respect to h.

Example 7.2 Let us solve the Cauchy problem (7.12) by using the Crank-Nicolson method with the same values of h as used in Example 7.1. As we can see, the results confirm that the error tends to zero with order 2:

```
>> y0=0; tspan=[0 1]; N=2; f=inline('cos(2*y)','t','y');
>> y='0.5*asin((exp(4*t)-1)./(exp(4*t)+1))';
>> for k=1:10
 [tt,u]=cranknic(f,tspan,y0,N);
 t=tt(end); e(k)=abs(u(end)-eval(y)); N=2*N; end
>> p=log(abs(e(1:end-1)./e(2:end)))/log(2); p(1:2:end)
    1.7940    1.9944    1.9997    2.0000    2.0000
```

7.4 Zero-stability

There is a concept of stability, called zero-stability, which guarantees that, in a fixed bounded interval, small perturbations of data yield bounded perturbations of the numerical solution when $h \to 0$.

More precisely, a numerical method for the approximation of problem (7.4), where $I = [t_0, T]$, is *zero-stable* if there exists $C > 0$ such that for all $\delta > 0$ and for any h sufficiently small

$$|z_n - u_n| \le C\delta, \ 0 \le n \le N_h, \tag{7.15}$$

where C is a constant which might depend on the length of the integration interval I, z_n is the solution that would be obtained by applying the numerical method at hand to a *perturbed* problem, and δ indicates the maximum size of the perturbation. Obviously, δ must be small enough to guarantee that the perturbed problem still has a unique solution on the interval of integration.

For instance, in the case of the forward Euler method, z_n satisfies

$$\begin{cases} z_{n+1} = z_n + h\left[f(t_n, z_n) + \rho_{n+1}\right], \\ z_0 = y_0 + \rho_0 \end{cases} \tag{7.16}$$

for $0 \leq n \leq N_h - 1$, under the assumption that $|\rho_n| \leq \delta$, $0 \leq n \leq N_h$.

For a consistent one-step method it can be proved that zero-stability is a consequence of the fact that f is Lipschitz-continuous with respect to its second argument (see, e.g. [QSS00]), in which case C depends on $\exp((T - t_0)L)$, where L is the Lipschitz constant. However, this is not necessarily true for other families of methods. Assume for instance that the numerical method can be written in the general form

$$u_{n+1} = \sum_{j=0}^{p} a_j u_{n-j} + h\sum_{j=0}^{p} b_j f_{n-j} + hb_{-1}f_{n+1}, \; n = p, p+1, \dots \tag{7.17}$$

for suitable coefficients $\{a_k\}$ and $\{b_k\}$ and $p > 0$. This is a linear *multistep method* and $p+1$ denotes the number of steps. We will see some example of multistep methods in Section 7.6. The polynomial

$$\pi(r) = r^{p+1} - \sum_{j=0}^{p} a_j r^{p-j}$$

is called the *first characteristic polynomial* associated with the numerical method (7.17), and we denote its roots by r_j, $j = 0, \dots, p$. The method (7.17) is zero-stable iff the following *root condition* is satisfied:

$$\begin{cases} |r_j| \leq 1 \text{ for all } j = 0, \dots, p; \\ \text{furthermore } \pi'(r_j) \neq 0 \text{ for those } j \text{ such that } |r_j| = 1. \end{cases} \tag{7.18}$$

For example, for the forward Euler method we have $p = 0$, $a_0 = 1$, $b_{-1} = 0$, $b_0 = 1$. For the backward Euler method we have $p = 0$, $a_0 = 1$, $b_{-1} = 1$, $b_0 = 0$ and for the Crank-Nicolson method we have $p = 0$, $a_0 = 1$, $b_{-1} = 1/2$, $b_0 = 1/2$. In all cases there is only one root of $\pi(r)$ which is equal to 1 and therefore all these methods are zero-stable.

The following property, known as Lax-Ritchmyer equivalence theorem, is most crucial in the theory of numerical methods (see, e.g., [IK66]), and highlights the fundamental role played by the property of zero-stability:

> *Any consistent method is convergent iff is zero-stable.*

See the Exercises 7.4-7.5.

7.5 Stability on unbounded intervals

In the previous section we considered the solution of the Cauchy problem on bounded intervals. In that context, the number N_h of subintervals depends on h and can become infinite only if h goes to zero.

On the other hand, there are several situations in which we wish to integrate the Cauchy problem on very large (virtually infinite) time intervals. In this case, even if h is fixed, N_h tends to infinity, and then results like (7.11) become meaningless as the right hand side of the inequality contains an unbounded quantity. We are interested in methods that are able to approximate the solution for arbitrarily long time-intervals, even with a step-size h relatively "large".

Unfortunately, the forward Euler method, which is very cheap to implement, does not enjoy this property. To see this, let us consider the following *model problem*

$$\begin{cases} y'(t) = \lambda y(t), & t \in (0, \infty), \\ y(0) = 1, \end{cases} \tag{7.19}$$

where λ is a negative real number. The exact solution is $y(t) = e^{\lambda t}$, which tends to 0 as t tends to infinity. Applying the forward Euler method to (7.19) we find that

$$u_0 = 1, \quad u_{n+1} = u_n(1 + \lambda h) = (1 + \lambda h)^{n+1}, \quad n \geq 0. \tag{7.20}$$

Thus $\lim_{n \to \infty} u_n = 0$ iff

$$-1 < 1 + h\lambda < 1, \quad i.e. \quad h < 2/|\lambda| \tag{7.21}$$

This condition expresses the requirement that, for *fixed* h, the numerical solution should reproduce the behavior of the exact solution when t_n tends to infinity. If $h > 2/|\lambda|$, then $\lim_{n \to \infty} |u_n| = +\infty$; thus (7.21) is a stability condition. The property that $\lim_{n \to \infty} u_n = 0$ is called *absolute stability*.

Example 7.3 Let us apply the forward Euler method to solve problem (7.19) with $\lambda = -1$. In that case we must have $h < 2$ for absolute stability. In Figure 7.3 we report the solutions obtained on the interval $[0, 30]$ for 3 different values of h: $h = 30/14(> 2)$, $h = 30/16(< 2)$ and $h = 1/2$. We can see that in the first two cases the numerical solution oscillates. However only in the first case (which violates the stability condition) the absolute value of the numerical solution does not vanish at infinity (and actually it diverges).

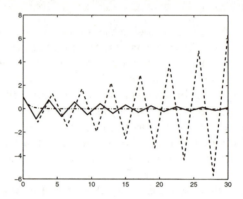

Fig. 7.3. Solutions of problem (7.19), with $\lambda = -1$, obtained by the forward Euler method, corresponding to $h = 30/14(> 2)$ (dashed line), $h = 30/16(< 2)$ (solid line) and $h = 1/2$ (dashed-dotted line)

Similar conclusions hold when λ is a negative function of t in (7.19). However in this case, $|\lambda|$ must be replaced by $\max_{t \in [0,\infty)} |\lambda(t)|$ in the stability condition. This condition could however be relaxed to one which is less strict by using a *variable step-size* h_n which accounts for the local behavior of $|\lambda(t)|$ in every interval (t_n, t_{n+1}).

In particular, the following *adaptive* forward Euler method could be used:

choose $u_0 = y_0$ and $h_0 = 2\alpha/|\lambda(t_0)|$; then

$$\text{for } n = 0, 1, \ldots, \text{ do}$$

$$t_{n+1} = t_n + h_n,$$

$$u_{n+1} = u_n + h_n \lambda(t_n) u_n, \tag{7.22}$$

$$h_{n+1} = 2\alpha/|\lambda(t_{n+1})|$$

where α is a constant which must be less than 1 in order to have an absolutely stable method.

For instance, consider the problem

$$y'(t) = -(10e^{-t} + 1)y(t), \qquad t \in (0, 10),$$

with $y(0) = 1$. Since $|\lambda(t)|$ is decreasing, the most restrictive condition for absolute stability of the forward Euler method is $h < h_0 = 2/|\lambda(0)| =$

2/11. In Figure 7.4, left, we compare the solution of the forward Euler method with that of the adaptive method (7.22) for three values of α. Note that, although every $\alpha < 1$ is admissible for stability purposes, to get an accurate solution requires choosing α sufficiently small. In Figure 7.4, right, we also plot the behaviour of h_n corresponding to the same values of α. This picture clearly shows that the sequence $\{h_n\}$ increases monotonically with n.

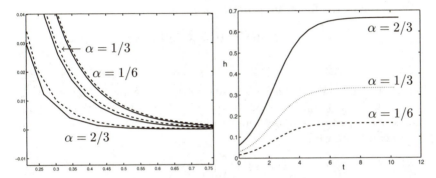

Fig. 7.4. Left: the numerical solution on the time interval $(0.2, 0.8)$ obtained by the forward Euler method with $h = \alpha h_0$ (dashed line) and by the adaptive variable stepping forward Euler method (7.22) (solid line) for three different values of α. Right: the behavior of the variable step-size h for the adaptive method (7.22)

In contrast to the forward Euler method, neither the backward Euler method nor the Crank-Nicolson method require limitations on h for absolute stability. In fact, with the backward Euler method we obtain $u_{n+1} = u_n + \lambda h u_{n+1}$ and therefore

$$u_{n+1} = \left(\frac{1}{1 - \lambda h}\right)^{n+1},$$

which tends to zero as $n \to \infty$ for *all values of $h > 0$*. Similarly, with the Crank-Nicolson method we obtain

$$u_{n+1} = \left[\left(1 + \frac{h\lambda}{2}\right) \bigg/ \left(1 - \frac{h\lambda}{2}\right)\right]^{n+1},$$

which still tends to zero as $n \to \infty$ for all possible values of $h > 0$. We can conclude that the forward Euler method is *conditionally absolutely stable*, while both the backward Euler and Crank-Nicolson methods are *unconditionally absolutely stable*. Methods which are unconditionally absolutely stable for all *complex* numbers λ in (7.19) which have real negative part are also called *A-stable*. Both Euler and Crank-Nicolson methods are A-stable. We point out that only implicit methods can be A-stable. This property makes implicit methods attractive in spite of being computationally more expensive than explicit methods.

7.5.1 Absolute stability controls perturbations

Consider now the following *generalized model problem*

$$\begin{cases} y'(t) = \lambda(t)y(t) + r(t), & t \in (0, +\infty), \\ y(0) = 1, \end{cases} \tag{7.23}$$

where λ and r are two continuous functions and $-\lambda_{max} \le \lambda(t) \le -\lambda_{min}$ with $0 < \lambda_{min} \le \lambda_{max} < +\infty$. In this case the exact solution does not necessarily tend to zero as t tends to infinity; for instance if both r and λ are constants we have

$$y(t) = \left(1 + \frac{r}{\lambda}\right)e^{\lambda t} - \frac{r}{\lambda}$$

whose limit when t tends to infinity is $-r/\lambda$. Thus, in general, it does not make sense to require a numerical method to be absolutely stable when applied to problem (7.23). However, we are going to show that a numerical method which is absolutely stable on the model problem (7.19), if applied to the generalized problem (7.23), guarantees that the perturbations are kept under control as t tends to infinity (possibly under a suitable constraint on the time-step h).

For the sake of simplicity we will confine our analysis to the forward Euler method; when applied to (7.23) it reads

$$\begin{cases} u_{n+1} = u_n + h(\lambda_n u_n + r_n), & n \ge 0, \\ u_0 = 1 \end{cases}$$

and its solution is (see Exercise 7.10)

$$u_n = u_0 \prod_{k=0}^{n-1}(1 + h\lambda_k) + h\sum_{k=0}^{n-1} r_k \prod_{j=k+1}^{n-1}(1 + h\lambda_j), \tag{7.24}$$

where $\lambda_k = \lambda(t_k)$ and $r_k = r(t_k)$. Let us consider the following "perturbed" method

$$\begin{cases} z_{n+1} = z_n + h(\lambda_n z_n + r_n + \rho_{n+1}), & n \ge 0, \\ z_0 = u_0 + \rho_0, \end{cases} \tag{7.25}$$

where $\rho_0, \rho_1, \ldots$ are given perturbations which are introduced at every time step. This is a simple model in which ρ_0 and ρ_{n+1}, respectively, account for the fact that neither u_0 nor r_n can be determined exactly. (Should we account for *all* roundoff errors which are actually introduced at any step, our perturbed model would be far more involved and diffi-cult to analyze.) The solution of (7.25) reads like (7.24) provided u_k is replaced by z_k and r_k by $r_k + \rho_{k+1}$, for all $k = 0, \ldots, n-1$. Then

$$z_n - u_n = \rho_0 \prod_{k=0}^{n-1}(1 + h\lambda_k) + h\sum_{k=0}^{n-1}\rho_{k+1}\prod_{j=k+1}^{n-1}(1 + h\lambda_j). \qquad (7.26)$$

It is worth noticing that this quantity does not depend on the function $r(t)$.

i. For the sake of exposition, let us consider first the special case where λ_k and ρ_k are two constants equal to λ and ρ, respectively. Assume that $h < h_0(\lambda) = 2/|\lambda|$, which is the condition on h that ensures the absolute stability of the forward Euler method applied to the model problem (7.19). Then, using the following identity for the geometric sum

$$\sum_{k=0}^{n-1}a^k = \frac{1 - a^n}{1 - a}, \qquad \text{if } |a| < 1, \qquad (7.27)$$

we obtain

$$z_n - u_n = \rho\left\{(1 + h\lambda)^n(1 + \frac{1}{\lambda}) - \frac{1}{\lambda}\right\}. \qquad (7.28)$$

It follows that the *perturbation error* $|z_n - u_n|$ satisfies (see Exercise 7.11)

$$|z_n - u_n| \leq \varphi(\lambda)|\rho|, \qquad (7.29)$$

with $\varphi(\lambda) = 1$ if $\lambda \leq -1$, while $\varphi(\lambda) = |1 + 2/\lambda|$ if $-1 \leq \lambda < 0$, and

$$\lim_{n\to\infty}|z_n - u_n| = \frac{\rho}{|\lambda|}.$$

For instance, Figure 7.5 corresponds to the case where $\rho = 0.1$, $\lambda = -2$ (left) and $\lambda = -0.5$ (right). In both cases we have taken $h = h_0(\lambda) - 0.01$. The conclusion that can be drawn is that the perturbation error is bounded by $|\rho|$ times a constant which is independent of n and h. Obvi-ously, the perturbation error blows up when n increases if the stability limit $h < h_0$ is violated.

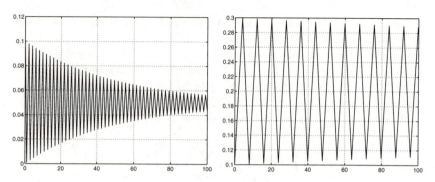

Fig. 7.5. The perturbation error when $\rho = 0.1$: $\lambda = -2$ (left) and $\lambda = -0.5$ (right). In both cases $h = h_0(\lambda) - 0.01$

ii. In the general case where λ and r are non constant, let us require h to satisfy the restriction $h < h_0(\lambda)$, where this time $h_0(\lambda) = 2/\lambda_{max}$. Then,

$$|1 + h\lambda_k| \leq a = a(h, \lambda_{min}, \lambda_{max}) = \max\{|1 - h\lambda_{min}|, |1 - h\lambda_{max}|\}.$$

Since $a < 1$, we can still use the identity (7.27) in (7.26) and obtain (see Exercise 7.11)

$$|z_n - u_n| \leq \rho_{max}\left(a^n + h\frac{1 - \delta(h)^n}{1 - \delta(h)}\right), \tag{7.30}$$

where $\rho_{max} = \max|\rho_k|$, $\delta(h) = |1 - h\lambda_{min}|$ if $h \leq h^*$ while $\delta(h) = |1 - h\lambda_{max}|$ if $h^* \leq h \leq 2/h_0(\lambda)$, having set $h^* = 2/(\lambda_{min} + \lambda_{max})$. When $h \leq h^*$, it follows that

$$|z_n - u_n| \leq \rho_{max}(a^n + 1/\lambda_{min}). \tag{7.31}$$

In particular,

$$\lim_{n\to\infty} |z_n - u_n| \leq \rho_{max}/\lambda_{min}, \tag{7.32}$$

from which we still conclude that the perturbation error is bounded by ρ_{max} times a constant which is independent of n and h (although the oscillations are no longer damped as in the previous case).

In fact, similar conclusion holds also when $h^* \leq h \leq 2/h_0(\lambda)$, although this does not follow from our upper bound (7.31) which is too pessimistic in this case. In Figure 7.6 we report the perturbation errors computed on the problem (7.23), where $\lambda(t) = -2 - \sin(t)$, $\rho_k = \rho(t_k)$, $\rho(t) = 0.1\sin(t)$ with $h < h^*$ (left) and with $h^* \leq h < h_0(\lambda)$ (right).

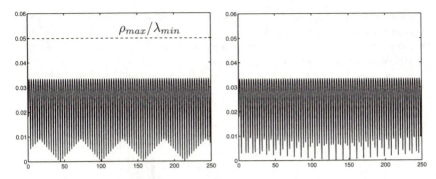

Fig. 7.6. The perturbation error when $\rho(t) = 0.1\sin(t)$ and $\lambda(t) = -2 - \sin(t)$: the step-size is $h = h^* - 0.01 = 0.39$ (left) and $h = h^* + 0.01$ (right)

iii. We consider now the general Cauchy problem (7.4). We claim that this problem can be related to the generalized model problem (7.23), in those cases where

$$-\lambda_{max} < \partial f/\partial y(t,y) < -\lambda_{min}, \forall t \geq 0, \ \forall y \in (-\infty, \infty),$$

for suitable values $\lambda_{min}, \lambda_{max} \in (0, +\infty)$. To this end, for every t in the generic interval (t_n, t_{n+1}), we subtract (7.5) from (7.16) to obtain the following equation for the perturbation error:

$$z_n - u_n = (z_{n-1} - u_{n-1}) + h\{f(t_{n-1}, z_{n-1}) - f(t_{n-1}, u_{n-1})\} + h\rho_n.$$

By applying the mean value theorem we obtain

$$f(t_{n-1}, z_{n-1}) - f(t_{n-1}, u_{n-1}) = \lambda_{n-1}(z_{n-1} - u_{n-1}),$$

where $\lambda_{n-1} = f_y(t_{n-1}, \xi_{n-1})$, having denoted $f_y = \partial f/\partial y$ and ξ_{n-1} being a suitable point in the interval whose endpoints are u_{n-1} and z_{n-1}. Thus

$$z_n - u_n = (1 + h\lambda_{n-1})(z_{n-1} - u_{n-1}) + h\rho_n.$$

By a recursive application of this formula we obtain the identity (7.26), from which we derive the same conclusions drawn in *ii.*, provided the stability restriction $0 < h < 2/\lambda_{max}$ holds.

Example 7.4 Let us consider the Cauchy problem

$$y'(t) = \arctan(3y) - 3y + t, \quad t > 0, \tag{7.33}$$

with $y(0) = 1$. Since $f_y = 3/(1 + 9y^2) - 3$ is negative, we can choose $\lambda_{max} = \max|f_y| = 3$ and set $h < 2/3$. Thus, we can expect that the perturbations on the forward Euler method are kept under control provided that $h < 2/3$. This is confirmed by the results which are reported in Figure 7.7. Note that in this example, taking $h = 2/3 + 0.01$ (thus violating the previous stability limit) the pertubation error blows up as t increases.

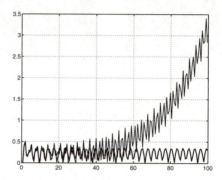

Fig. 7.7. The perturbation errors when $\rho(t) = \sin(t)$ with $h = 2/\lambda_{max} - 0.01$ (thick line) and $h = 2/\lambda_{max} + 0.01$ (thin line) for the Cauchy problem (7.33)

Example 7.5 We seek a limit on h that guarantees stability for the forward Euler method applied to approximate the Cauchy problem

$$y' = 1 - y^2, \quad t > 0, \tag{7.34}$$

with $y(0) = \dfrac{e-1}{e+1}$. The exact solution is $y(t) = (e^{2t+1} - 1)/(e^{2t+1} + 1)$ and $f_y = -2y$. Since $f_y \in (-2, -0.9)$ for all $t > 0$, we can take h less than $h_0 = 1$. In Figure 7.8, left, we report the solutions obtained on the interval $(0, 35)$ with $h = 0.95$ (thick line) and $h = 1.05$ (thin line). In both cases the solution oscillates, but remains bounded. Moreover in the first case, which satisfies the stability constraint, the oscillations are damped and the numerical solution tends to the exact one as t increases. In Figure 7.8, right, we report the perturbation errors corresponding to $\rho(t) = \sin(t)$ with $h = 0.95$ (thick line) and $h = h^* + 0.1$ (thin line). In both cases the perturbation errors remain bounded; moreover in the former case the upper bound (7.32) is satisfied.

In those cases where no information on y is available, finding the value $\lambda_{max} = \max|f_y|$ is not a simple matter. A more heuristic approach could be pursued in these situations, by adopting a variable stepping procedure. Precisely, one could take $t_{n+1} = t_n + h_n$, where

$$h_n < 2\frac{\alpha}{|f_y(t_n, u_n)|},$$

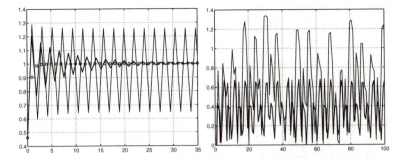

Fig. 7.8. On the left, numerical solutions of problem (7.34) obtained by the forward Euler method with $h = 20/19$ (thin line) and $h = 20/21$ (thick line). The values of the exact solution are indicated by circles. On the right, perturbation errors corresponding to $\rho(t) = \sin(t)$ with $h = 0.95$ (thick line) and $h = h^*$ (thin line)

for suitable values of α strictly less than 1. Note that the denominator depends on the value u_n which is known. In Figure 7.9 we report the perturbation errors corresponding to the Example 7.5 for two different values of α.

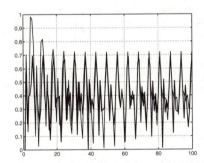

Fig. 7.9. The perturbation errors corresponding to $\rho(t) = \sin(t)$ with $\alpha = 0.8$ (thick line) and $\alpha = 0.9$ (thin line) for the Example 7.5, using the adaptive strategy

Remark 7.4 The previous analysis can be carried out also for other kind of one-step methods, in particular for the backward Euler and Crank-Nicolson methods. For these methods which are A-stable, the same conclusions about the perturbation error can be drawn without requiring any limitation on the time-step. In fact, in the previous analysis one should replace each term $1 + h\lambda_k$ by $(1 - h\lambda_k)^{-1}$ in the backward Euler case and by $(1 + h\lambda_k/2)/(1 - h\lambda_k/2)$ in the Crank-Nicolson case.

1. An absolutely stable method is one which generates a solution u_n of the model problem (7.19) which tends to zero as t_n tends to infinity;

2. a method is said *A-stable* if it is absolutely stable for any possible choice of the time-step h (otherwise a method is called conditionally stable, and h should be lower than a constant depending on λ);

3. when an absolutely stable method is applied to a generalized model problem (like (7.23)), the perturbation error (that is the absolute value of the difference between the perturbed and unperturbed solution) is uniformly bounded (with respect to h). In short we can say that absolutely stable method keeps the perturbation controlled;

4. the analysis of absolute stability for the linear model problem can be exploited to find stability conditions on the time-step when considering the nonlinear Cauchy problem (7.4) with a function f satisfying $\partial f / \partial y < 0$. In that case the stability restriction requires the step-size to be chosen as a function of $\partial f / \partial y$. Precisely, the new integration interval $[t_n, t_{n+1}]$ is chosen in such a way that $h_n = t_{n+1} - t_n$ satisfies $h_n < 2/|\partial f(t_n, u_n)/\partial y|$.

See the Exercises 7.6-7.14.

7.6   High order methods

All methods presented so far are elementary examples of one-step methods. More sophisticated schemes, which allow the achievement of a higher order of accuracy, are the *Runge-Kutta methods* and the *multistep methods*. Runge-Kutta methods are still one-step methods; however, they involve several evaluations of the function $f(t, y)$ on every interval $[t_n, t_{n+1}]$. One of the most celebrated Runge-Kutta methods reads

$$u_{n+1} = u_n + \frac{h}{6}(K_1 + 2K_2 + 2K_3 + K_4)$$

where

$$K_1 = f_n,$$
$$K_2 = f(t_n + \tfrac{h}{2}, u_n + \tfrac{h}{2}K_1),$$
$$K_3 = f(t_n + \tfrac{h}{2}, u_n + \tfrac{h}{2}K_2),$$
$$K_4 = f(t_{n+1}, u_n + hK_3).$$

It is an explicit method of order four with respect to h; at each time step, it involves four new evaluations of the function f. Other Runge-Kutta methods, either explicit or implicit, can be constructed with arbitrary order.

They stand at the base of a family of MATLAB programs whose names contain the root ode followed by numbers and letters. In particular, ode45 is based on a pair of explicit Runge-Kutta methods (the so-called Dormand-Prince pair) of order 4 and 5, respectively. ode23 is the implementation of another pair of explicit Runge-Kutta methods (the Bogacki and Shampine pair). In these methods the integration step varies in order to guarantee that the error is below a given tolerance (in these programs the default scalar relative error tolerance RelTol is equal to 10^{-3}). The program ode23tb is an implementation of an implicit Runge-Kutta formula whose first stage is the trapezoidal rule, while the second stage is a backward differentiation formula of order two (see later).

Multistep methods (see (7.17)) achieve a high order of accuracy by involving several values $u_n, u_{n-1}, \ldots$, for the determination of u_{n+1}. They can be derived for instance by applying the fundamental theorem of integration (which we recalled in Section 1.4.3) to the Cauchy problem (7.4), obtaining

$$y_{n+1} = y_n + \int_{t_n}^{t_{n+1}} f(t, y(t))\, dt, \tag{7.35}$$

and then approximating the integral by a quadrature formula which involves the interpolant of f at a suitable set of nodes. A notable example of multistep method is the three-steps, third order (explicit) Adams-Bashforth formula (AB3)

$$u_{n+1} = u_n + \frac{h}{12}\left(23 f_n - 16 f_{n-1} + 5 f_{n-2}\right) \tag{7.36}$$

which is obtained by replacing f in (7.35) by its interpolating polynomial of degree 2 at the nodes t_{n-2}, t_{n-1}, t_n. Another important example is the three-steps, fourth order (implicit) Adams-Moulton formula (AM4)

$$u_{n+1} = u_n + \frac{h}{24} \left(9f_{n+1} + 19f_n - 5f_{n-1} + f_{n-2}\right) \qquad (7.37)$$

which is obtained by replacing f in (7.35) by its interpolating polynomial of degree 3 at the nodes $t_{n-2}, t_{n-1}, t_n, t_{n+1}$.

Another family of multistep methods can be obtained by writing the differential equation at time t_{n+1} and replacing $y'(t_{n+1})$ by a one-sided incremental ratio of high order. An instance is provided by the formula

$$u_{n+1} = \frac{18}{11}u_n - \frac{9}{11}u_{n-1} + \frac{2}{11}u_{n-2} + \frac{6h}{11}f_{n+1} \qquad (7.38)$$

which is known as the three-steps, third order (implicit) *backward difference formula* (BDF3).

All these methods are consistent and zero-stable. Indeed, the first characteristic polynomial of both (7.36) and (7.37) is $\pi(r) = r^3 - r^2$ and its roots are $r_0 = 1$, $r_1 = r_2 = 0$, while the first characteristic polynomial of (7.38) is $\pi(r) = r^3 - 18/11r^2 + 9/11r - 2/11$ and its roots are $r_0 = 1$, $r_1 = 0.3182 + 0.2839i$, $r_2 = 0.3182 - 0.2839i$, where i is the imaginary unit. In all cases, the root condition (7.18) is satisfied.

Moreover, when applied to the model problem (7.19), AB3 is absolutely stable if $h < 0.54/|\lambda|$ while AM4 is absolutely stable if $h < 3/|\lambda|$. The method BDF3 is unconditionally absolutely stable for all real negative λ. However, this is no longer true when $\lambda \in \mathbb{C}$ with negative real part. In other words, BDF3 fails to be A-stable. More generally, there is no multistep *A-stable* method of order strictly greater than two.

Multistep methods are implemented in several MATLAB programs, for instance in ode15s.

ode15s

7.7 The predictor-corrector methods

In Section 7.2 it was pointed out that implicit methods yield at each step a nonlinear problem for the unknown value u_{n+1}. For its solution we can use one of the methods introduced in Chapter 2, or else apply the function fzero as we have done with the Programs 13 and 14.

Alternatively, we can carry out fixed point iterations at every time-step. As an example, for the Crank-Nicolson method (7.13), for $k =$

$0, 1, \ldots$, we compute until convergence

$$u_{n+1}^{(k+1)} = u_n + \frac{h}{2}\left[f_n + f(t_{n+1}, u_{n+1}^{(k)})\right].$$

It can be proved that if the initial guess $u_{n+1}^{(0)}$ is chosen conveniently, a single iteration suffices in order to obtain a numerical solution $u_{n+1}^{(1)}$ whose accuracy is of the same order as the solution u_{n+1} of the original implicit method. More precisely, if the original implicit method has order p, then the initial guess $u_{n+1}^{(0)}$ must be generated by an explicit method of order (at least) $p - 1$.

For instance, if we use the first order (explicit) forward Euler method to initialize the Crank-Nicolson method, we get the *Heun method* (also called the *improved Euler method*), which is a second order explicit Runge-Kutta method:

$$\begin{aligned} u_{n+1}^* &= u_n + hf_n, \\ u_{n+1} &= u_n + \frac{h}{2}\left[f_n + f(t_{n+1}, u_{n+1}^*)\right] \end{aligned} \qquad (7.39)$$

The explicit step is called a *predictor*, whereas the implicit one is called a *corrector*. Another example combines the (AB3) method (7.36) as predictor with the (AM4) method (7.37) as corrector. These kinds of methods are therefore called *predictor-corrector* methods. They enjoy the order of accuracy of the corrector method. However, being explicit, they undergo a stability restriction which is typically the same as that of the predictor method. Thus they are not adequate to integrate a Cauchy problem on unbounded intervals.

In Program 15 we implement a general predictor-corrector method. The strings `predictor` and `corrector` identify the type of method that is chosen. For instance, if we use the functions `eeonestep` and `cnonestep`, which are defined in Program 16, we can call `predcor` as follows:

$>>$ [t,u]=predcor(t0,y0,T,N,f,'eeonestep','cnonestep');

and obtain the Heun method.

Program 15 - predcor: predictor-corrector method

```
function [t,u]=predcor(odefun,tspan,y,Nh,pred,corr,varargin)
%PREDCOR Solve differential equations using a predictor-corrector method
%   [T,Y]=PREDCOR(ODEFUN,TSPAN,Y0,NH,PRED,CORR) with TSPAN =
```

```
%  [T0 TFINAL] integrates the system of differential equations y' = f(t,y) from time
%   T0 to TFINAL with initial conditions Y0 using a general predictor corrector
%   method on an equispaced grid of NH intervals. Function  ODEFUN(T,Y)
%   must return a column vector corresponding to f(t,y). Each row in the
%   solution array Y corresponds to a time returned in the column vector T.
%   Functions PRED and CORR identify the type of method that is chosen.
%  [T,Y] = PREDCOR(ODEFUN,TSPAN,Y0,NH,PRED,CORR,P1,P2,...) passes
%   the additional parameters P1,P2,... to the functions ODEFUN, PRED and
%  CORR as ODEFUN(T,Y,P1,P2...), PRED(T,Y,P1,P2...), CORR(T,Y,P1,P2...).
h=(tspan(2)-tspan(1))/Nh;  tt=[tspan(1):h:tspan(2)];
u=y; [n,m]=size(u); if n ¡ m, u=u'; end
for t=tt(1:end-1)
   y = u(:,end); fn = feval(odefun,t,y,varargin{:});
   upre = feval(predictor,t,y,h,fn);
   ucor = feval(corrector,t+h,y,upre,h,odefun,fn,varargin{:});
   u = [u, ucor];
end
t = tt;
```

Program 16 - onestep: one step of forward Euler (eeonestep),
one step of backward Euler (eionestep), one step of Crank-
Nicolson (cnonestep)

```
function [u]=feonestep(t,y,h,f)
u = y + h*f;
return

function [u]=beonestep(t,u,y,h,f,fn,varargin)
u = u + h*feval(f,t,y,varargin{:});
return

function [u]=cnonestep(t,u,y,h,f,fn,varargin)
u = u + 0.5*h*(feval(f,t,y,varargin{:})+fn);
return
```

ode113   The MATLAB program ode113 implements an Adams Moulton Bash-
forth scheme with variable step-size.

See the Exercises 7.15-7.18.

7.8 Systems of differential equations

Let us consider the following system of first-order ordinary differential equations whose unknowns are $y_1(t), \ldots, y_m(t)$:

$$y_1' = f_1(t, y_1, \ldots, y_m),$$

$$\vdots$$

$$y_m' = f_m(t, y_1, \ldots, y_m),$$

where $t \in (t_0, T]$, with the initial conditions

$$y_1(t_0) = y_{0,1}, \quad \ldots \quad y_m(t_0) = y_{0,m}.$$

For its solution we could apply to each individual equation one of the methods previously introduced for a scalar problem. For instance, the n-th step of the forward Euler method would read

$$\begin{cases} u_{n+1,1} = u_{n,1} + h f_1(t_n, u_{n,1}, \ldots, u_{n,m}), \\ \vdots \\ u_{n+1,m} = u_{n,m} + h f_1(t_n, u_{n,1}, \ldots, u_{n,m}). \end{cases}$$

By writing the system in vector form $\mathbf{y}'(t) = \mathbf{F}(t, \mathbf{y}(t))$, with obvious choice of notation, the extension of the methods previously developed for the case of a single equation to the vector case is straightforward. For instance, the method

$$\mathbf{u}_{n+1} = \vartheta \mathbf{F}(t_{n+1}, \mathbf{u}_{n+1}) + (1 - \vartheta)\mathbf{F}(t_n, \mathbf{u}_n), \qquad n \geq 0,$$

with $\mathbf{u}_0 = \mathbf{y}_0$, $0 \leq \vartheta \leq 1$, is the vector form of the forward Euler method if $\vartheta = 0$, the backward Euler method if $\vartheta = 1$ and the Crank-Nicolson method if $\vartheta = 1/2$.

Example 7.6 Let us apply the forward Euler method to solve the Lotka-Volterra equations (7.3) with $C_1 = C_2 = 1$, $b_1 = b_2 = 0$ and $d_1 = d_2 = 1$. In order to use Program 12 for a *system* of ordinary differential equations, let us create a function f which contains the component of the vector function **F**, which we save in the file f.m. For our specific system we have:

```
C1=1; C2=1; d1=1; d2=1; b1=0; b2=0;
yy(1)=C1*y(1)*(1-b1*y(1)-d2*y(2));   % first equation
y(2)=-C2*y(2)*(1-b2*y(2)-d1*y(1));   % second equation
y(1)=yy(1);
```

Now we execute Program 12 with the following instructions

```
[t,u]=feuler('f',[0,10],[2 2],2000);
plot(t,u); figure(2); plot(u(:,1),u(:,2));
```

They correspond to solving the Lotka-Volterra system on the time-interval $[0, 10]$ with a time-step $h = 0.005$.

The graph in Figure 7.10, left, represents the time evolution of the two components of the solution. Note that they are periodic with period 2π. The second graph in Figure 7.10, right, shows the trajectory issuing from the initial value in the so-called *phase plane*, that is, the cartesian plane whose coordinate axes are y_1 and y_2. This trajectory is confined within a bounded region of the (y_1, y_2) plane. If we start from the point $(1.2, 1.2)$, the trajectory would stay in an even smaller region surrounding the point $(1, 1)$. This can be explained as follows. Our differential system admits 2 *points of equilibrium* at which $y_1' = 0$ and $y_2' = 0$, and one of them is precisely $(1, 1)$ (the other being $(0, 0)$). Actually, they are obtained by solving the nonlinear system

$$y_1' = y_1 - y_1 y_2 = 0,$$

$$y_2' = -y_2 + y_2 y_1 = 0.$$

If the initial data coincide with one of these points, the solution remains constant in time. Moreover, while $(0, 0)$ is an unstable equilibrium point, $(1, 1)$ is stable, that is, all trajectories issuing from a point near $(1, 1)$ stay bounded in the phase plane.

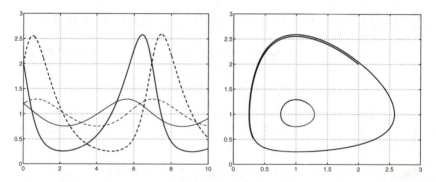

Fig. 7.10. Numerical solutions of system (7.3). On the left, we represent y_1 and y_2 on the time interval $(0, 10)$, the solid line refers to y_1, the dashed line to y_2. Two different initial data are considered: $(2, 2)$ (thick lines) and $(1.2, 1.2)$ (thin lines). On the right, we report the corresponding trajectories in the phase plane

When we use an explicit method, the step-size h should undergo a stability restriction similar to the one encountered in Section 7.5. In the case when the real part of the eigenvalues λ_k of the Jacobian $A(t) = [\partial \mathbf{F}/\partial \mathbf{y}](t, \mathbf{y})$ of $\mathbf{F}$ are all negative, we can set $\lambda = -\max_t \rho(A(t))$, where $\rho(A(t))$ is the spectral radius of $A(t)$. This λ is a candidate to replace the one entering in the stability conditions (such as, $e.g.$, (7.21)) that were derived for the scalar Cauchy problem.

Remark 7.5 The MATLAB programs (ode23, ode45, ...) that we have mentioned before can be used also for the solution of systems of ordinary differential equations. The synthax is odeXX('f',[t0 tf],y0), where y0 is the vector of the initial conditions, f is a function to be specifed by the user and odeXX is one of the methods available in MATLAB.

Now we consider the case of an ordinary differential equation of order m

$$y^{(m)}(t) = f(t, y, y', \dots, y^{(m-1)}) \tag{7.40}$$

for $t \in (t_0, T]$, whose solution (when existing) is a family of functions defined up to m arbitrary constants. The latter can be fixed by prescribing m initial conditions

$$y(t_0) = y_0, \quad y'(t_0) = y_1, \quad \dots \quad y^{(m-1)}(t_0) = y_m.$$

Setting

$$w_1(t) = y(t), \quad w_2(t) = y'(t), \quad \dots \quad w_m(t) = y^{(m-1)}(t),$$

the equation (7.40) can be transformed into a first-order system of m differential equations

$$w_1' = w_2,$$

$$w_2' = w_3,$$

$$\vdots$$

$$w_{m-1}' = w_m,$$

$$w_m' = f(t, w_1, \dots, w_m),$$

with initial conditions

$$w_1(t_0) = y_0, \quad w_2(t_0) = y_1, \quad \dots \quad w_m(t_0) = y_{m-1}.$$

Thus we can always approximate the solution of a differential equation of order $m > 1$ by resorting to the equivalent system of m first-order equations, and then applying to this system a convenient discretization method.

Example 7.7 Consider the circuit of Problem 7.3 and suppose that $L(i_1) = L$ is constant and that $R_1 = R_2 = R$. In this case v can be obtained by solving the following system of two differential equations:

$$\begin{cases} v' = w, \\ w' = -\dfrac{L + RC^2}{LCR}w - \dfrac{2}{LC}v + \dfrac{e}{LC}, \end{cases} \quad (7.41)$$

with initial conditions $v(0) = 0$, $w(0) = 0$. The system has been obtained from the second-order differential equation

$$LC\frac{d^2v}{dt^2} + \left(\frac{L}{R_2} + R_1C\right)\frac{dv}{dt} + \left(\frac{R_1}{R_2} + 1\right)v = e.$$

We set $L = 0.1$ Henry, $C = 10^{-3}$ Farad, $R = 10$ Ohm and $e = 5$ Volt, where Henry, Farad, Ohm and Volt are respectively the unit measure of inductance, capacitance, resistance and voltage. Now we apply the forward Euler method with $h = 0.001$ seconds in the time interval $[0, 0.1]$, by the Program 12:

```
>> [t,u]=feuler('fsys',[0,0.1],[0 0],1000);
```

where **fsys** is contained in the file **fsys.m**:

```
function y=fsys(t,y)
L=0.1; C=1.e-03; R=10; e=5;
yy(1)=y(2);
y(2)=-(L/R+R*C)/(L*C)*y(2)-2/(L*C)*y(1)+e/(L*C);
y(1)=yy(1);
```

In Figure 7.11 we report the approximated values of v and w. As expected, $v(t)$ tends to $e/2 = 2.5$ Volt for large t. In this case the real part of the eigenvalues of $A(t) = [\partial F/\partial y](t, y)$ is negative and λ can be set equal to -141.4214. Then a condition for absolute stability is to take $h < 2/|\lambda| = 0.0282$.

See the Exercises 7.19-7.20.

7.9   What we haven't told you

For a complete derivation of the whole family of the Runge-Kutta methods we refer to [But87], [Lam91] and [QSS00], Chapter 11.

For derivation and analysis of multistep methods we refer to [Arn73] and [Lam91].

We have not mentioned the so-called *stiff* problems, e.g. those associated with differential systems for which the Jacobian of **F** is an ill-conditioned matrix. Their numerical solution requires *ad hoc* implicit

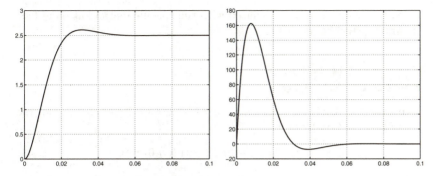

Fig. 7.11. Numerical solution of system (7.41). The potential drop $v(t)$ is reported on the left, its derivative w on the right

methods, since otherwise unreasonably small time-steps would be required (see [QSS00], Chapter 11, [Lam91] and [DB02]). The MATLAB programs ode15s, ode23s and ode23tb are suited for stiff problems. Consider for instance the Van Der Pol system

$$\begin{cases} y_1' = y_2, \\ y_2' = 1000(1 - y_1^2)y_2 - y_1, \\ y_1(0) = 2, \qquad y_2(0) = 0. \end{cases} \tag{7.42}$$

Using the program ode45 on the interval $(0, 3000)$ would require an unreasonably large CPU time due to the tiny time-step which is imposed by stability purpose. On the contrary, ode15s provides an accurate solution with "only" 592 time-steps (see Figure 7.12).

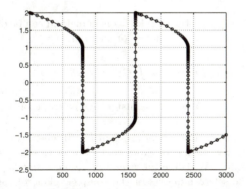

Fig. 7.12. The first component y_1 of the solution of (7.42) computed using the program ode15s. Note the non-uniform distribution of the integration nodes

We note that a variable step-size is adopted, which guarantees both stability and the desired accuracy. Its choice is based on suitable error indicators which can be associated with the specific method at hand (see [Lam91], [SR97] and [DB02]).

7.10 Exercises

Exercise 7.1 Apply the backward Euler and forward Euler methods for the solution of the Cauchy problem

$$y' = \sin(t) + y, \quad t \in (0, 1], \quad \text{with } y(0) = 0, \tag{7.43}$$

and verify that both converge with order 1.

Exercise 7.2 Consider the Cauchy problem

$$y' = -te^{-y}, \quad t \in (0, 1], \quad \text{with } y(0) = 0. \tag{7.44}$$

Apply the forward Euler method with $h = 1/100$ and estimate the number of exact significant digits of the approximate solution at $t = 1$ (use the property that the value of the exact solution is included between -1 and 0).

Exercise 7.3 The backward Euler method applied to problem (7.44) requires at each step the solution of the non-linear equation: $u_{n+1} = u_n - ht_{n+1}e^{-u_{n+1}} = \phi(u_{n+1})$. The solution u_{n+1} can be obtained by the following fixed-point iteration: for $k = 0, 1, \ldots$, compute $u_{n+1}^{(k+1)} = \phi(u_{n+1}^{(k)})$, with $u_{n+1}^{(0)} = u_n$. Find under which restriction on h these iterations converge.

Exercise 7.4 Repeat Exercise 7.1 for the Crank-Nicolson method.

Exercise 7.5 Verify that the Crank-Nicolson method can be derived from the following integral form of the Cauchy problem (7.4)

$$y(t) - y_0 = \int_{t_0}^{t} f(\tau, y(\tau))d\tau$$

provided that the integral is approximated by the trapezoidal formula (4.17).

Exercise 7.6 Consider the model problem (7.19) where λ is now a complex number with negative real part. Note that even in this case $u(t)$ tends to 0 as t tends to infinity. We define the *region of absolute stability* of a numerical method to be the set of numbers of the complex plane of the form $z = h\lambda$ for which the method is absolutely stable. Determine the region of absolute stability associated with the forward Euler method.

Exercise 7.7 Solve the model problem (7.19) with $\lambda = -1+i$ by the forward Euler method and find the values of h for which we have absolute stability.

Exercise 7.8 Show that the Heun method defined in (7.39) is consistent. Write a MATLAB program to implement it for the solution of the Cauchy problem (7.43) and verify experimentally that the method has order of convergence equal to 2 with respect to h.

Exercise 7.9 Prove that the Heun method (7.39) is absolutely stable if $-2 \leq h\lambda \leq 0$ where λ is real and negative.

Exercise 7.10 Prove formula (7.24).

Exercise 7.11 Prove the inequality (7.29).

Exercise 7.12 Prove the inequality (7.30).

Exercise 7.13 Verify the consistency of the following method (which is an explicit Runge-Kutta method of order 3)

$$u_{n+1} = u_n + \frac{1}{6}(k_1 + 4k_2 + k_3), \quad k_1 = hf(t_n, u_n),$$
$$k_2 = hf(t_n + \tfrac{h}{2}, u_n + \tfrac{k_1}{2}), \quad k_3 = hf(t_{n+1}, u_n + 2k_2 - k_1). \tag{7.45}$$

Write a MATLAB program to implement it for the solution of the Cauchy problem (7.43) and verify experimentally that the method has order of convergence equal to 3 with respect to h. The methods (7.39) and (7.45) stand at the base of the MATLAB program ode23 for the solution of ordinary differential equations.

Exercise 7.14 Prove that the method (7.45) is absolutely stable if $-2.5 \leq h\lambda \leq 0$ where λ is real and negative.

Exercise 7.15 The *modified Euler method* is defined as follows:

$$u_{n+1}^* = u_n + hf(t_n, u_n), \quad u_{n+1} = u_n + hf(t_{n+1}, u_{n+1}^*). \tag{7.46}$$

Find under which condition on h this method is absolutely stable.

Exercise 7.16 Solve equation (7.1) by the Crank-Nicolson method and the Heun method when the body in question is a cube with side equal to 1 m and mass equal to 1 Kg. Assume that $T_0 = 180K$, $T_e = 200K$, $\gamma = 0.5$ and $C = 100 J/(Kg/K)$. Compare the results obtained by using $h = 20$ and $h = 10$, for t ranging from 0 to 200 seconds.

Exercise 7.17 Use MATLAB to compute the region of absolute stability of the Heun method.

Exercise 7.18 Solve the Cauchy problem (7.12) by the Heun method and verify its order.

Exercise 7.19 The displacement $x(t)$ of a vibrating system represented by a body of a given weight and a spring, subjected to a resistive force proportional to the velocity, is described by the second-order differential equation $x'' + 5x' + 6x = 0$. Solve it by the Heun method assuming that $x(0) = 1$ and $x'(0) = 0$, for $t \in [0, 5]$.

Exercise 7.20 The motion of a frictionless Foucault pendulum is described by the two equations

$$x'' - 2\omega \sin(\Psi)y' + k^2 x = 0, \quad y'' + 2\omega \cos(\Psi)x' + k^2 y = 0,$$

where Ψ is the latitude of the place where the pendulum is located, $\omega = 7.29 \cdot 10^{-5}$ sec^{-1} is the angular velocity of the Earth, $k = \sqrt{g/l}$ with $g = 9.8$ m/sec^2 and l is the length of the pendulum. Apply the forward Euler method to compute $x = x(t)$ and $y = y(t)$ for t ranging between 0 and 300 seconds and $\Psi = \pi/4$.

8. Numerical methods for boundary-value problems

Boundary-value problems are differential problems set in an interval (a, b) of the real line or in an open multidimensional region $\Omega \subset \mathbb{R}^d$ ($d = 2, 3$) for which the value of the unknown solution (or its derivatives) is prescribed at the end-points a and b of the interval, or on the boundary $\partial \Omega$ of the multidimensional region.

In the multidimensional case the differential equation will involve *partial derivatives* of the exact solution with respect to the space coordinates. Some examples of boundary-value problems are reported below.

1. *Poisson equation*:

$$-u''(x) = f(x), \quad x \in (a, b), \tag{8.1}$$

or (in several dimensions)

$$-\Delta u(\mathbf{x}) = f(\mathbf{x}), \quad \mathbf{x} \in \Omega, \tag{8.2}$$

where f is a given function and Δ is the so-called *Laplace operator*:

$$\Delta u = \sum_{i=1}^{d} \frac{\partial^2 u}{\partial x_i^2}.$$

The symbol $\partial \cdot / \partial x_i$ denotes partial derivative with respect to the x_i variable, that is, for every point $\mathbf{x}^0$

$$\frac{\partial u}{\partial x_i}(\mathbf{x}^0) = \lim_{h \to 0} \frac{u(\mathbf{x}^0 + h\mathbf{e}_i) - u(\mathbf{x}^0)}{h}, \tag{8.3}$$

where $\mathbf{e}_i$ is i-th unitary vector of $\mathbb{R}^d$.

2. *Heat equation*:

$$\frac{\partial u(x,t)}{\partial t} - \mu \frac{\partial^2 u(x,t)}{\partial x^2} = f(x,t), \quad x \in (a,b), \ t > 0,$$

or (in several dimensions)

$$\frac{\partial u(\mathbf{x},t)}{\partial t} - \mu \Delta u(\mathbf{x},t) = f(\mathbf{x},t), \quad \mathbf{x} \in \Omega, \ t > 0,$$

where t is the time variable, $\mu > 0$ is a given coefficient representing the thermal conductivity, and f is again a given function.

3. *Wave equation*:

$$\frac{\partial^2 u(x,t)}{\partial t^2} - c \frac{\partial^2 u(x,t)}{\partial x^2} = 0, \quad x \in (a,b), \ t > 0,$$

or (in several dimensions)

$$\frac{\partial^2 u(\mathbf{x},t)}{\partial t^2} - c \Delta u(\mathbf{x},t) = 0, \quad \mathbf{x} \in \Omega, \ t > 0,$$

where c is a positive constant.

We will restrict our attention to equations like (8.1) and (8.2). Equations depending on time, like the heat equation and the wave equation, are called initial-boundary value problems. In that case an initial condition $u = u_0$ at $t = 0$ needs to be prescribed as well. For initial-boundary value problems and for more general partial differential equations, the reader is referred for instance to [QV94], [EEHJ96] or [Lan03].

Problem 8.1 (Thermodynamics) If we are interested in the temperature distribution in a square Ω with side of length L, we can compute the net energy variation in each direction in an infinitesimal square with side of length $l \ll L$. We have

$$J(\mathbf{x}) - J(\mathbf{x} + l\mathbf{e}_i) = -l\mathbf{e}_i \frac{\partial J}{\partial x_i}(\mathbf{x}),$$

where $J(\mathbf{x})$ represents the energy transfer for unit time. The Fourier law states that J is proportional to the variation of the temperature T, and then

$$J(\mathbf{x}) - J(\mathbf{x} + l\mathbf{e}_i) = -l\mathbf{e}_i \frac{\partial}{\partial x_i} \left(kl \frac{\partial T}{\partial x_i} \right) = kl^2 \frac{\partial^2 T}{\partial x_i^2},$$

where k is a positive constant expressing the thermal conductivity coefficient. At equilibrium the sum of energy variations must vanish, and therefore

$$\Delta T(\mathbf{x}) = \sum_{i=1}^{d} \frac{\partial^2 T}{\partial x_i^2}(\mathbf{x}) = 0 \quad \mathbf{x} \in \Omega,$$

which is a Poisson problem with $f = 0$.   •

Problem 8.2 (Hydrogeology) The study of filtration in groundwater can lead, in some cases, to an equation like (8.2). Consider a portion Ω occupied by a porous medium (like ground or clay). According to the Darcy law, the water velocity filtration $\mathbf{q} = (q_1, q_2, q_3)^T$ is equal to the variation of the water level ϕ in the medium, precisely

$$\mathbf{q} = -K(\partial\phi/\partial x_1, \partial\phi/\partial x_2, \partial\phi/\partial x_3)^T, \qquad (8.4)$$

where K is the constant hydraulic conductivity of the porous medium. Assume that the fluid density is constant; then the mass conservation principle yields the equation $\mathrm{div}\mathbf{q} = 0$, where $\mathrm{div}\mathbf{q}$ is the *divergence* of the vector $\mathbf{q}$ and is defined as

$$\mathrm{div}\mathbf{q} = \sum_{i=1}^{3} \frac{\partial q_i}{\partial x_i}.$$

Thanks to (8.4) we therefore find that ϕ satisfies the Poisson problem $\Delta\phi = 0$ (see Exercise 8.9).   •

The previous Poisson equation (8.2) admits an infinite number of solutions. With the aim of obtaining a unique solution we must impose suitable conditions on the boundary $\partial\Omega$ of Ω. One possibility is to prescribe the value of u on $\partial\Omega$, that is

$$u(\mathbf{x}) = g(\mathbf{x}) \text{ for } \mathbf{x} \in \partial\Omega, \qquad (8.5)$$

where g is a given function.

Problem (8.2), supplemented by the boundary condition (8.5), is a *Dirichlet boundary-value problem*, and is precisely the problem that we will face in next section. An alternative to (8.5) is to prescribe a value for the partial derivative of u with respect to the normal direction to the boundary $\partial\Omega$, in which case we will get a *Neumann boundary-value problem*.

It can be proven that if f and g are two continuous functions and the region Ω is regular enough, then the Dirichlet boundary-value problem has a unique solution (while the solution of the Neumann boundary-value problem is unique up to an additive constant).

In the one-dimensional case, the Dirichlet boundary-value problem takes the following form: being given two constants α and β and a function $f = f(x)$, find a function $u = u(x)$ which satisfies

$$\begin{array}{ll} -u''(x) = f(x) & \text{for } x \in (a, b), \\[2mm] u(a) = \alpha, & u(b) = \beta \end{array} \qquad (8.6)$$

Performing double integration it is easily seen that if $f \in C^0([a, b])$, the solution u exists and is unique; moreover it belongs to $C^2([a, b])$.

Although (8.6) is an ordinary differential problem, it cannot be cast in the form of a Cauchy problem for ordinary differential equations since the value of u is prescribed at two different points.

The numerical methods which are used for its solution are based on the same principles which form the basis of the approximation of multi-dimensional boundary-value problems. This is the reason why in Section 8.1 we will make a digression on the numerical solution of problem (8.6).

8.1 Approximation of boundary-value problems

We introduce on $[a, b]$ a partition into intervals $I_j = [x_j, x_{j+1}]$ for $j = 0, \ldots, N$ with $x_0 = a$ and $x_{N+1} = b$. We assume for simplicity that all intervals have the same length h.

8.1.1 Approximation by finite differences

The differential equation must be satisfied in particular at any point x_j (which we call *nodes* from now on) internal to (a, b), that is

$$-u''(x_j) = f(x_j), \quad j = 1, \ldots, N.$$

We can approximate this set of N equations by replacing the second derivative with a suitable finite difference as we have done in Chapter 4 for the first derivatives. In particular, we observe that if $u : [a, b] \to \mathbb{R}$ is a sufficiently smooth function in a neighborhood of a generic point $\bar{x} \in (a, b)$, then the quantity

$$\delta^2 u(\bar{x}) = \frac{u(\bar{x} + h) - 2u(\bar{x}) + u(\bar{x} - h)}{h^2} \tag{8.7}$$

provides an approximation to $u''(\bar{x})$ of order 2 with respect to h (see Exercise 8.3). This suggests the use of the following approximation to problem (8.6): find $\{u_j\}_{j=1}^N$ such that

$$-\frac{u_{j+1} - 2u_j + u_{j-1}}{h^2} = f(x_j), \quad j = 1, \ldots, N, \tag{8.8}$$

with $u_0 = \alpha$ and $u_{N+1} = \beta$. Equations (8.8) provide a linear system

$$\mathbf{A}\mathbf{u}_h = h^2 \mathbf{f}, \tag{8.9}$$

where $\mathbf{u}_h = (u_1, \ldots, u_N)^T$ is the vector of unknowns, $\mathbf{f} = (f(x_1) + \alpha/h^2, f(x_2), \ldots, f(x_{N-1}), f(x_N) + \beta/h^2)^T$, and A is the tridiagonal matrix

$$\begin{bmatrix} 2 & -1 & 0 & \cdots & & 0 \\ -1 & 2 & \ddots & & & \vdots \\ 0 & \ddots & \ddots & -1 & 0 & \\ \vdots & & -1 & 2 & -1 \\ 0 & \cdots & 0 & -1 & 2 \end{bmatrix} . \qquad (8.10)$$

This system admits a unique solution since A is symmetric and positive definite (see Exercise 8.1). Since A is tridiagonal, this system can be solved by the Thomas algorithm introduced in Section 5.4. We note however that, for small values of h (and thus for large values of N), A is ill-conditioned. Indeed, $K(A) = \lambda_{max}(A)/\lambda_{min}(A) = Ch^{-2}$, for a suitable constant C independent of h (see Exercise 8.2). Consequently, the numerical solution of system (8.9), by either direct or iterative methods, requires special care. In particular, when using iterative methods a suitable preconditioner ought to be employed.

It is possible to prove (see, *e.g.*, [QSS00], Chapter 12) that if $f \in C^2([a, b])$ then

$$\max_{j=0,\ldots,N+1} |u(x_j) - u_j| \le \frac{h^2}{96} \max_{x \in [a,b]} |f''(x)|$$

that is, the finite difference method (8.8) converges with order 2 with respect to h.

In Program 17 we solve the boundary-value problem

$$\begin{cases} -u''(x) + \delta u'(x) + \gamma u(x) = f(x) & \text{for } x \in (a, b), \\ u(a) = \alpha & u(b) = \beta, \end{cases} \qquad (8.11)$$

which is a generalization of problem (8.6). For this problem the finite difference method, which generalizes (8.8), reads:

$$\begin{cases} -\dfrac{u_{j+1} - 2u_j + u_{j-1}}{h^2} + \delta \dfrac{u_{j+1} - u_{j-1}}{2h} + \gamma u_j = f(x_j), & j = 1, \ldots, N, \\ u_0 = \alpha, & u_{N+1} = \beta. \end{cases}$$

The input parameters of Program 17 are the end-points a and b of the interval, the number N of internal nodes, the constant coefficients δ and

γ and the string bvpfun. Finally, ua and ub represent the values that the solution should attain at x=a and x=b. Output parameters are the vector of nodes x and the computed solution uh.

Program 17 - bvp: Approximation of a 2-point boundary-value problem by the finite difference method

```
function [x,uh]=bvp(a,b,N,delta,gamma,bvpfun,ua,ub,varargin)
%BVP Solve two-point boundary value problems.
%   [X,UH]=BVP(A,B,N,DELTA,GAMMA,BVPFUN,UA,UB) solves with the
%   centered finite difference method the boundary-value
%   problem
%      -D(DU/DX)/DX+DELTA*DU/DX+GAMMA*U=BVPFUN
%   on the interval (A,B) with boundary conditions U(A)=UA
%   and U(B)=UB.
%   BVPFUN can be an inline function.
h = (b-a)/(N+1);
z = linspace(a,b,N+2);
e = ones(N,1);
A = spdiags([-e-0.5*h*delta 2*e+gamma*h^2 -e+0.5*h*delta], -1:1, N, N);
x = z(2:end-1);
f = h^2*feval(bvpfun,x,varargin{:});
f=f';   f(1) = f(1) + ua;   f(end) = f(end) + ub;
uh = A\f;
uh=[ua; uh; ub];
x = z;
```

8.1.2   Approximation by finite elements

The *finite element method* represents an alternative to the previous finite difference method. It is derived from a suitable reformulation of problem (8.6).

Let us multiply both sides of (8.6) by a generic function v. Integrating the corresponding equality on the interval (a, b), we obtain

$$-\int_a^b u''(x)v(x) \; dx = \int_a^b f(x)v(x) \; dx. \qquad (8.12)$$

If we assume that $v \in C^1([a, b])$ and use integration by parts we obtain

$$\int_a^b u'(x)v'(x) \; dx - [u'(x)v(x)]_a^b = \int_a^b f(x)v(x) \; dx.$$

By making the further assumption that v vanishes at the end-points $x = a$ and $x = b$, problem (8.6) becomes: find u such that $u(a) = \alpha$, $u(b) = \beta$ and

$$\int_a^b u'(x)v'(x)\, dx = \int_a^b f(x)v(x)\, dx \tag{8.13}$$

for each $v \in C^1([a, b])$ such that $v(a) = v(b) = 0$. This is called *weak formulation* of problem (8.6). (Indeed, the test functions v can be less regular than $C^1([a, b])$, see, *e.g.* [QSS00].)

Its finite element approximation is defined as follows:

find $u_h \in V_h$ such that $u_h(a) = \alpha$, $u_h(b) = \beta$ and

$$\sum_{j=0}^{N} \int_{x_j}^{x_{j+1}} u_h'(x)v_h'(x)\, dx = \int_a^b f(x)v_h(x)\, dx, \qquad \forall v_h \in V_h^0 \tag{8.14}$$

where

$$V_h = \left\{ v_h \in C^0([a, b]) : \ v_{h|I_j} \in \mathbb{P}_1, j = 0, \dots, N \right\},$$

i.e. V_h is the space of continuous functions on (a, b) whose restrictions on every sub-interval I_j are linear polynomials. Moreover, V_h^0 is the subspace of V_h of those functions vanishing at the end-points a and b. V_h is called space of finite elements of degree 1.

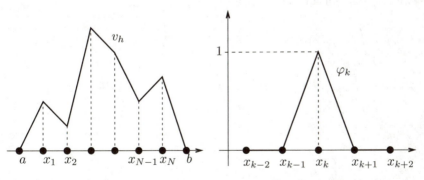

Fig. 8.1. To the left, a generic function $v_h \in V_h^0$. To the right, the basis function of V_h^0 associated with the k-th node

The functions in V_h^0 are piecewise linear polynomials (see Figure 8.1, left). In particular, every function v_h of V_h^0 admits the representation

$$v_h(x) = \sum_{j=1}^{N} v_h(x_j) \varphi_j(x),$$

where

$$\varphi_j(x) = \begin{cases} \dfrac{x - x_{j-1}}{x_j - x_{j-1}} & \text{if } x \in I_{j-1}, \\ \dfrac{x - x_{j+1}}{x_j - x_{j+1}} & \text{if } x \in I_j, \\ 0 & \text{otherwise,} \end{cases}$$

for $j = 1, \ldots, N$. Thus, φ_k is null at every node x_j except at x_k where $\varphi_k(x_k) = 1$ (see Figure 8.1, right). The functions φ_j, $j = 1, \ldots, N$ are called *shape functions* and provide a basis for the vector space V_h^0.

Consequently, we can limit ourselves to fulfill (8.14) only for the shape functions φ_j, $j = 1, \ldots, N$. By exploiting the fact that φ_j vanishes outside the intervals I_{j-1} and I_j, from (8.14) we obtain

$$\int_{I_{j-1} \cup I_j} u_h'(x) \varphi_j'(x)\, dx = \int_{I_{j-1} \cup I_j} f(x) \varphi_j(x)\, dx, \quad \forall j = 1, \ldots, N. \quad (8.15)$$

On the other hand, we can write $u_h(x) = \sum_{j=1}^{N} u_j \varphi_j(x) + \alpha \varphi_0(x) + \beta \varphi_{N+1}(x)$, where $u_j = u_h(x_j)$, $\varphi_0(x) = (a + h - x)/h$ for $a \leq x \leq a+h$, and $\varphi_{N+1}(x) = (x - b + h)/h$ for $b - h \leq x \leq b$, while both $\varphi_0(x)$ and $\varphi_{N+1}(x)$ are zero otherwise. By substituting this expression in (8.15), we find that for all $j = 1, \ldots, N$

$$u_{j-1} \int_{I_{j-1}} \varphi_{j-1}'(x) \varphi_j'(x)\, dx + u_j \int_{I_{j-1} \cup I_j} \varphi_j'(x) \varphi_j'(x)\, dx$$

$$+ u_{j+1} \int_{I_j} \varphi_{j+1}'(x) \varphi_j'(x)\, dx = \int_{I_{j-1} \cup I_j} f(x) \varphi_j(x)\, dx + B_{1,j} + B_{N,j}$$

where

$$B_{1,j} = \begin{cases} -\alpha \displaystyle\int_{I_0} \varphi_0'(x) \varphi_1'(x)\, dx = -\dfrac{\alpha}{x_1 - a} & \text{if } j = 1, \\ 0 \text{ otherwise,} \end{cases}$$

while

$$B_{N,j} = \begin{cases} -\beta \displaystyle\int_{I_N} \varphi'_{N+1}(x)\varphi'_j(x)\, dx = -\dfrac{\beta}{b-x_N} \text{if } j = N, \\ 0 \text{ otherwise.} \end{cases}$$

In the special case where all intervals have the same length h, then $\varphi'_{j-1} = -1/h$ in I_{j-1}, $\varphi'_j = 1/h$ in I_{j-1} and $\varphi'_j = -1/h$ in I_j, $\varphi'_{j+1} = 1/h$ in I_j. Consequently, we obtain for $j = 1, \ldots, N$

$$-u_{j-1} + 2u_j - u_{j+1} = h \int_{I_{j-1} \cup I_j} f(x)\varphi_j(x)\, dx + A_{0,j} + B_{0,j}.$$

This linear system has the same matrix as the finite difference system (8.9), but a different right-hand side (and a different solution too, in spite of coincidence of notation). Finite difference and finite element solutions share however the same accuracy with respect to h when the nodal maximum error is computed.

Obviously the finite element approach can be generalized to problems like (8.11) (also in the case when δ and γ depend on x). A further generalization consists of using piecewise polynomials of degree greater than 1, allowing the achievement of higher convergence orders. In these cases, the finite element matrix does not coincide anymore with that of finite differences, and the convergence order is greater than when using piecewise linear polynomials.

See Exercises 8.1-8.8.

8.2 Finite differences in 2 dimensions

Let us consider a partial differential equation, for instance equation (8.2), in a two-dimensional region Ω.

The idea behind finite differences relies on approximating the partial derivatives that are present in the PDE again by incremental ratios computed on a suitable grid (called the computational grid) made of a finite number of nodes. Then the solution u of the PDE will be approximated only at these nodes.

The first step therefore consists of introducing a computational grid. Assume for simplicity that Ω is the rectangle $(a, b) \times (c, d)$. Let us introduce a partition of $[a, b]$ in subintervals (x_k, x_{k+1}) for $k = 0, \ldots, N_x$, with $x_0 = a$ and $x_{N_x+1} = b$. Let us denote by $\Delta_x = \{x_0, \ldots, x_{N_x+1}\}$ the set of end points of such intervals and by $h_x = \max\limits_{k=0,\dots,N_x} (x_{k+1} - x_k)$ their maximum length.

In a similar manner we introduce a discretization of the y-axis $\Delta_y = \{y_0, \dots, y_{N_y+1}\}$ with $y_0 = c$ and $y_{N_y+1} = d$. The cartesian product $\Delta_h = \Delta_x \times \Delta_y$ provides the computational grid on Ω (see Figure 8.2), and $h = \max\{h_x, h_y\}$ is a characteristic measure of the grid-size. We are looking for values $u_{i,j}$ which approximate $u(x_i, y_j)$. We will assume for the sake of simplicity that the nodes be uniformly spaced, that is, $x_i = x_0 + i h_x$ for $i = 0, \dots, N_x+1$ and $y_j = y_0 + j h_y$ for $j = 0, \dots, N_y+1$.

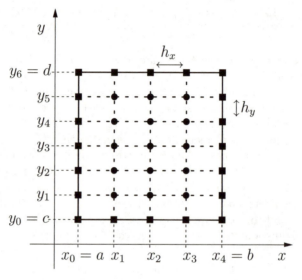

Fig. 8.2. The computational grid Δ_h with only 15 internal nodes on a rectangular domain

The second order partial derivatives of a function can be approximated by a suitable incremental ratio, as we did for ordinary derivatives. In the case of a function of 2 variables, we define the following incremental ratios:

$$\delta_x^2 u_{i,j} = \frac{u_{i-1,j} - 2u_{i,j} + u_{i+1,j}}{h_x^2}, \quad \delta_y^2 u_{i,j} = \frac{u_{i,j-1} - 2u_{i,j} + u_{i,j+1}}{h_y^2}. \quad (8.16)$$

They are second order accurate with respect to h_x and h_y, respectively, for the approximation of $\partial^2 u/\partial x^2$ and $\partial^2 u/\partial y^2$ at the node (x_i, y_j). If we replace the second order partial derivatives of u with the formula (8.16), by requiring that the PDE is satisfied at all internal nodes of Δ_h, we obtain the following set of equations:

$$-(\delta_x^2 u_{i,j} + \delta_y^2 u_{i,j}) = f_{i,j}, \quad i = 1, \ldots, N_x, \ j = 1, \ldots, N_y. \tag{8.17}$$

We have set $f_{i,j} = f(x_i, y_j)$. We must add the equations that enforce the Dirichlet data at the boundary, which are

$$u_{i,j} = g_{i,j} \quad \forall i, j \text{ such that } (x_i, y_j) \in \partial \Delta_h, \tag{8.18}$$

where $\partial \Delta_h$ indicates the set of nodes belonging to the boundary $\partial \Omega$ of Ω. These nodes are indicated by small squares in Figure 8.2. If we make the further assumption that the computational grid is uniform in both cartesian directions, that is, $h_x = h_y = h$, instead of (8.17) we obtain

$$-\frac{1}{h^2}(u_{i-1,j} + u_{i,j-1} - 4u_{i,j} + u_{i,j+1} + u_{i+1,j}) = f_{i,j},$$
$$i = 1, \ldots, N_x, \ j = 1, \ldots, N_y \tag{8.19}$$

The system given by equations (8.19) (or (8.17)) and (8.18) allows the computation of the nodal values $u_{i,j}$ at all nodes of Δ_h. For every fixed pair of indices i and j, equation (8.19) involves 5 unknown nodal values as we can see in Figure 8.3. For that reason this finite difference scheme is called *the 5 point scheme* for the Laplace operator. We note that the unknowns associated with the boundary nodes can be eliminated using (8.18) (or (8.17)), and therefore (8.19) involves only $N = N_x N_y$ unknowns.

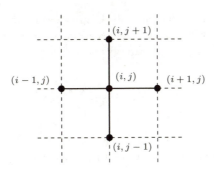

Fig. 8.3. The stencil of the 5 point scheme for the Laplace operator

The resulting system can be written in a more interesting form if we adopt the *lexicographic* order according to which the nodes (and, correspondingly, the unknown components) are numbered by proceeding from left to right, from the top to the bottom. We obtain a system of the form (8.9), with a matrix $A \in \mathbb{R}^{N \times N}$ which takes the following block tridiagonal form:

$$
A = \begin{bmatrix}
T & D & 0 & \cdots & 0 \\
D & T & \ddots & & \vdots \\
0 & \ddots & \ddots & D & 0 \\
\vdots & & D & T & D \\
0 & \cdots & 0 & D & T
\end{bmatrix}.
$$

There are N_y rows and N_y columns, and every entry (denoted by a capital letter) consists of a $N_x \times N_x$ matrix. In particular, $D \in \mathbb{R}^{N_x \times N_x}$ is a diagonal matrix whose diagonal entries are $-1/h_y^2$, while $T \in \mathbb{R}^{N_x \times N_x}$ is a symmetric tridiagonal matrix

$$
T = \begin{bmatrix}
\frac{2}{h_x^2} + \frac{2}{h_y^2} & -\frac{1}{h_x^2} & 0 & \cdots & & 0 \\
-\frac{1}{h_x^2} & \frac{2}{h_x^2} + \frac{2}{h_y^2} & \ddots & & & \vdots \\
0 & \ddots & \ddots & -\frac{1}{h_x^2} & & 0 \\
\vdots & & -\frac{1}{h_x^2} & \frac{2}{h_x^2} + \frac{2}{h_y^2} & -\frac{1}{h_x^2} \\
0 & \cdots & & 0 & -\frac{1}{h_x^2} & \frac{2}{h_x^2} + \frac{2}{h_y^2}
\end{bmatrix}.
$$

A is symmetric since all diagonal blocks are symmetric. It is also positive definite, that is $\mathbf{v}^T A \mathbf{v} > 0 \; \forall \mathbf{v} \in \mathbb{R}^N$, $\mathbf{v} \neq \mathbf{0}$. Actually, by partitioning $\mathbf{v}$ in N_y vectors $\mathbf{v}_i$ of length N_x we obtain

$$
\mathbf{v}^T A \mathbf{v} = \sum_{k=1}^{N_y} \mathbf{v}_k^T T \mathbf{v}_k - \frac{2}{h_y^2} \sum_{k=1}^{N_y-1} \mathbf{v}_k^T \mathbf{v}_{k+1}. \tag{8.20}
$$

We can write $T = 2/h_y^2 I + 1/h_x^2 K$ where K is the (symmetric and positive definite) matrix given in (8.10). Consequently, (8.20) becomes

$$
(\mathbf{v}_1^T K \mathbf{v}_1 + \mathbf{v}_2^T K \mathbf{v}_2 + \ldots + \mathbf{v}_{N_y}^T K \mathbf{v}_{N_y})/h_x^2
$$

which is a strictly positive real number since K is positive definite and at least one vector $\mathbf{v}_i$ is non null.

Having proven that A is non-singular we can conclude that the finite difference system admits a unique solution $\mathbf{u}_h$.

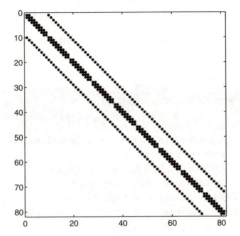

Fig. 8.4. Pattern of the matrix associated with the 5-point scheme using the lexicographic ordering of the unknowns

The matrix A is *sparse* since the number of non-null elements is much smaller that of null elements. (In Figure 8.4 we report the structure of the matrix corresponding to a uniform grid of 11×11 nodes. This picture was obtained by using the command spy(A). It can be noted that the **spy** only nonzero elements lie on 5 diagonals).

A sparse matrix can be stored in a special mode that allows an easy access to the nonzero elements and preserves only these elements. The command sparse that we use in the Program 18 will serve this purpose. **sparse** Since A is symmetric and positive definite, the associated system can be solved efficiently by either direct or iterative methods, as illustrated in Chapter 5. Finally, it is worth pointing out that A shares with its one-dimensional analog the property of being ill-conditioned: indeed, its condition number grows like h^{-2} as h tends to zero, where $h = \max(h_x, h_y)$.

In the Program 18 we construct and solve the system (8.17)-(8.18). The input parameters a, b, c and d denote the corners of the rectangular domain $\Omega = (a, c) \times (b, d)$, while nx and ny denote the values of N_x and N_y (the case $N_x \neq N_y$ is admitted). Finally, the two strings fun and bound represent the right-hand side $f = f(x, y)$ (otherwise called the source term) and the boundary data $g = g(x, y)$. The output is a two-dimensional array u whose i, j-th entry is the nodal value $u_{i,j}$. The numerical solution can be visualized by the command mesh(x,y,u). The **mesh** (optional) string uex represents the exact solution of the original problem for those cases (of theoretical interest) where this solution is known. In such cases the output parameter error contains the nodal relative error between the exact and numerical solution, which is computed as follows:

$$\texttt{error} = \max_{i,j}|u(x_i,y_j) - u_{i,j}|/\max_{i,j}|u(x_i,y_j)|.$$

Program 18 - poissonfd: Approximation of the Poisson problem with Dirichlet data by the five-point finite difference method

```
function [u,x,y,error]=poissonfd(a,c,b,d,nx,ny,fun,bound,uex,varargin)
%POISSONFD two-dimensional Poisson solver
%    [U,X,Y]=POISSONFD(A,C,B,D,NX,NY,FUN,BOUND) solves by
%    the 5-points finite difference scheme the problem
%    -LAPL(U) = FUN in the rectangle (A,C)X(B,D) with
%    Dirichlet boundary conditions U(X,Y)=BOUND(X,Y) for any
%    (X,Y) on the boundary of the rectangle.
%
%    [U,X,Y,ERROR]=POISSONFD(A,C,B,D,NX,NY,FUN,BOUND,UEX) computes
%    also the maximum nodal error ERROR with respect to the exact
%    solution UEX.
%    FUN, BOUND and UEX can be online functions.
if nargin == 8
    uex = inline('0','x','y');
end
nx=nx+1;         ny=ny+1;
hx = (b-a)/nx;   hy = (d-c)/ny;
nx1 = nx+1;      hx2 = hx^2;
hy2 = hy^2;
kii = 2/hx2+2/hy2;    kix = -1/hx2;    kiy = -1/hy2;
dim = (nx+1)*(ny+1);
K = speye(dim,dim);
rhs = zeros(dim,1);
y = c;
for m = 2:ny
    x = a; y = y + hy;
    for n = 2:nx
        i = n+(m-1)*(nx+1);
        x = x + hx;
        rhs(i) = feval(fun,x,y,varargin{:});
        K(i,i) = kii;
        K(i,i-1) = kix;
        K(i,i+1) = kix;
        K(i,i+nx1) = kiy;
        K(i,i-nx1) = kiy;
    end
end
rhs1 = zeros(dim,1);
```

```
x = [a:hx:b];
rhs1(1:nx1) = feval(bound,x,c,varargin{:});
rhs1(dim-nx:dim) = feval(bound,x,d,varargin{:});
y = [c:hy:d];
rhs1(1:nx1:dim-nx) = feval(bound,a,y,varargin{:});
rhs1(nx1:nx1:dim) = feval(bound,b,y,varargin{:});
rhs = rhs - K*rhs1;
nbound = [[1:nx1],[dim-nx:dim],[1:nx1:dim-nx],[nx1:nx1:dim]];
ninternal = setdiff([1:dim],nbound);
K = K(ninternal,ninternal);
rhs = rhs(ninternal);
utemp = K\rhs;
uh = rhs1;
uh (ninternal) = utemp;
k = 1; y = c;
for j = 1:ny+1
    x = a;
    for i = 1:nx1
        u(i,j) = uh(k);
        k = k + 1;
        ue(i,j) = feval(uex,x,y,varargin{:});
        x = x + hx;
    end
    y = y + hy;
end
x = [a:hx:b];
y = [c:hy:d];
if nargout == 4 & nargin == 8
        warning('Exact solution not available');
        error = [ ];
    else
        error = max(max(abs(u-ue)))/max(max(abs(ue)));
end, end
return
```

Example 8.1 The transverse displacement u of an elastic membrane from a reference plane $\Omega = (0,1)^2$ under a load whose intensity is $f(x,y) = 8\pi^2 \sin(2\pi x) \cos(2\pi y)$ satisfies a Poisson problem like (8.2) in the domain Ω. The Dirichlet value of the displacement is prescribed on $\partial\Omega$ as follows: $g = 0$ on the sides $x = 0$ and $x = 1$, and $g(x,0) = g(x,1) = \sin(2\pi x)$, $0 < x < 1$. This problem admits the exact solution $u(x,y) = \sin(2\pi x)\cos(2\pi y)$. In Figure 8.5 we show the numerical solution obtained by the five-point finite difference scheme on a uniform grid. Two different values of h have been used: $h = 1/10$ (left) and $h = 1/20$ (right). When h decreases the numerical solution improves, and actually the nodal relative error is 0.0292 for $h = 1/10$ and 0.0081 for $h = 1/20$.

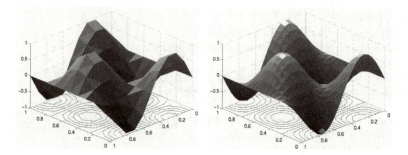

Fig. 8.5. Transverse displacement of an elastic membrane computed on two uniform grids. On the horizontal plane we report the isolines of the numerical solution. The triangular partition of Ω only serves the purpose of the visualization of the results

Also the finite element method can be easily extended to the two-dimensional case. To this end the problem (8.2) must be reformulated in an integral form and the partition of the interval (a, b) in one dimension must be replaced by a decomposition of Ω by polygons (typically, triangles) called *elements*. The shape function φ_k will still be a continuous function, whose restriction on each element is a polynomial of degree 1 on each element, which is equal to 1 at the k-th vertex (or node) of the triangulation and 0 at all other vertices. For its implementation one can

 pde use the MATLAB toolbox `pde`.

8.2.1 Consistency and convergence

In the previous section we have shown that the solution of the finite difference problem exists and is unique. Now we investigate the approximation error. We will assume for simplicity that $h_x = h_y = h$. If

$$\max_{i,j} |u(x_i, y_j) - u_{i,j}| \to 0 \quad \text{as } h \to 0,$$

the method is called convergent.

As we have already pointed out, consistency is a necessary condition for convergence. A method is consistent if the residual that is obtained when the exact solution is plugged into the numerical scheme tends to zero when h tends to zero. If we consider the five point finite difference scheme, at every internal node (x_i, y_j) of Δ_h we define

$$\tau_h(x_i, y_j) = -f(x_i, y_j)$$

$$- \frac{u(x_{i-1}, y_j) + u(x_i, y_{j-1}) - 4u(x_i, y_j) + u(x_i, y_{j+1}) + u(x_{i+1}, y_j)}{h^2}.$$

This is the *local truncation error* at the node (x_i, y_j). By (8.2) we obtain

$$\tau_h(x_i, y_j) = \left\{ \frac{\partial^2 u}{\partial x^2}(x_i, y_j) - \frac{u(x_{i-1}, y_j) - 2u(x_i, y_j) + u(x_{i+1}, y_j)}{h^2} \right\}$$
$$+ \left\{ \frac{\partial^2 u}{\partial y^2}(x_i, y_j) - \frac{u(x_i, y_{j-1}) - 2u(x_i, y_j) + u(x_i, y_{j+1})}{h^2} \right\}.$$

Thanks to the analysis that was carried out in Section 8.2 we can conclude that both terms vanish as h tends to 0. Thus

$$\lim_{h \to 0} \tau_h(x_i, y_j) = 0, \quad \forall (x_i, y_j) \in \Delta_h \setminus \partial \Delta_h,$$

that is, the five-point method is consistent. It is also convergent, as stated in the following Proposition (for its proof, see, *e.g.*, [IK66]):

Proposition 8.1 *Assume that the exact solution* $u \in C^4(\bar{\Omega})$, *i.e. all its partial derivatives up to the 4th order are continuous in the closed domain* $\bar{\Omega}$. *Then there exists a positive constant* C *such that*

$$\max_{i,j} |u(x_i, y_j) - u_{i,j}| \leq CMh^2$$

where M *is the maximum absolute value attained by the fourth order derivatives of* u *in* $\bar{\Omega}$.

Example 8.2 Let us verify that the 5 point scheme applied to solve the Poisson problem of Example 8.1 converges with order 2 with respect to h. We start from $h = 1/4$ and, then we halve subsequently the value of h, until $h = 1/64$, through the following instructions:

```
>> a=0;b=1;c=0;d=1;
>> f=inline('8*pi^2*sin(2*pi*x).*cos(2*pi*y)','x','y');
>> g=inline('sin(2*pi*x).*cos(2*pi*y)','x','y'); uex=g; nx=4; ny=4;
>> for n=1:5
    [u,x,y,error(n)]=poissonfd(a,c,b,d,nx,ny,f,g,uex); nx = 2*nx; ny = 2*ny;
end
```

The vector containing the error is

```
>> format short e; error
    1.3565e-01   4.3393e-02   1.2308e-02   3.2775e-03   8.4557e-04
```

As we can verify using the following commands

```
>> log(abs(error(1:end-1)./error(2:end)))/log(2)
    1.6443e+00   1.8179e+00   1.9089e+00   1.9546e+00
```

this error decreases as h^2.

Let us summarize

1. Boundary-value problems are differential equations set in a spatial domain $\Omega \subset \mathbb{R}^d$ (which is an interval if $d = 1$) that require information on the solution on the domain boundary;

2. finite difference approximations are based on the discretization of the given differential equation at selected points (called nodes) where derivatives are replaced by finite difference formulae;

3. the finite difference method provides a nodal vector whose components converge to the corresponding nodal values of the exact solution quadratically with respect to the grid-size;

4. the finite element method is based on a suitable integral reformulation of the original differential equation, then on the assumption that the approximate solution is a piecewise polynomial;

5. matrices arising from both finite difference and finite element approximations are sparse and ill-conditioned.

See Exercises 8.9-8.10.

8.3   What we haven't told you

We could simply say that we have told you almost nothing, since the field of numerical analysis which is devoted to the numerical approximation of partial differential equations is so broad and multifaceted to deserve an entire monograph simply for addressing the most essential concepts (see, *e.g.*, [TW98], [EEHJ96]).

We would like to mention that the finite element method is nowadays probably the most widely diffused method for the numerical solution of partial differential equations (see, *e.g.*, [QV94], [Bra97], [BS01]). As already mentioned the MATLAB toolbox `pdetool` allows the solution of a broad family of partial differential equations by the linear finite element method.

Other popular techniques are the spectral methods (see, [CHQZ88], [Fun92], [BM92], [KS99]) and the finite volume method (see, [Krö98], [Hir88] and [LeV02]).

8.4 Exercises

Exercise 8.1 Verify that matrix (8.10) is positive definite.

Exercise 8.2 Verify that the eigenvalues of the matrix (8.10) are

$$\lambda_j = 2(1 - \cos(j\theta)), \quad j = 1, \dots, N,$$

while the corresponding eigenvectors are

$$\mathbf{q}_j = (\sin(j\theta), \sin(2j\theta), \dots, \sin(nj\theta))^T,$$

where $\theta = \pi/(n+1)$. Deduce that $K(A)$ is proportional to h^{-2}.

Exercise 8.3 Prove that the quantity (8.7) provides a second order approximation of $u''(\bar{x})$ with respect to h.

Exercise 8.4 Compute the matrix and the right hand side of the numerical scheme that we have proposed to approximate problem (8.11).

Exercise 8.5 Use the finite difference method to approximate the boundary-value problem

$$\begin{cases} -u'' + \dfrac{k}{T}u = \dfrac{w}{T} & \text{in } (0,1), \\ u(0) = u(1) = 0, \end{cases}$$

where $u = u(x)$ represents the vertical displacement of a string of length 1, subject to a transverse load of intensity w per unit length. T is the tension and k is the elastic coefficient of the string. For the case in which $w = 1+\sin(4\pi x)$, $T = 1$ and $k = 0.1$, compute the solution corresponding to $h = 1/i$, $i = 10, 20, 40$, and deduce the order of accuracy of the method.

Exercise 8.6 We consider problem (8.11) on the interval $(0,1)$ with $\gamma = 0$, $f = 0$, $\alpha = 0$ and $\beta = 1$. Using the Program 17 find the maximum value h_{crit} of h for which the numerical solution is monotone (as is the exact solution) when $\delta = 100$. What happens if $\delta = 1000$? Suggest an empirical formula for $h_{crit}(\delta)$ as a function of δ, and verify it for several values of δ.

Exercise 8.7 Use the finite difference method to solve problem (8.11) in the case where the following *Neumann* boundary conditions are prescribed at the endpoints

$$u'(a) = \alpha, \quad u'(b) = \beta.$$

Use the formulae given in (4.10) to discretize $u'(a)$ and $u'(b)$.

Exercise 8.8 Verify that, when using a uniform grid, the right hand side of the system associated with the centered finite difference scheme coincides with that of the finite element scheme provided that the composite trapezoidal formula is used to compute the integrals on the elements I_{k-1} and I_k.

Exercise 8.9 Verify that $\mathrm{div}\nabla\phi = \Delta\phi$, where ∇ is the gradient operator that associates to a function u the vector whose components are the first order partial derivatives of u.

Exercise 8.10 Consider a square plate whose side is 20 cm and thermal conductivity is $k = 0.2$ cal/sec·cm·C. Denote by $Q = 5$ cal/cm³·sec the heat production rate per unit area. The temperature $T = T(x,y)$ of the plate satisfies the equation $-\Delta T = Q/k$. Assuming that T is null on three sides of the plate and is equal to 1 on the fourth side, determine the temperature T at the center of the plate.

9. Solutions of the exercises

Chapter 1

Solution 1.1 Only the numbers of the form $\pm 0.1a_1 \cdot 2^t$ with $a_1 = 0, 1$ and $t = \pm 2, \pm 1, 0$ belong to the set $\mathbb{F}(2, 2, -2, 2)$. For a given exponent, we can represent in this set only the two numbers 0.10 and 0.11, and their opposites. Consequently, the number of elements belonging to $\mathbb{F}(2, 2, -2, 2)$ is 20. Finally, $\epsilon_M = 1/2$.

Solution 1.2 For any fixed exponent, each of the digits $a_2, \ldots, a_t$ can assume β different values, while a_1 can assume only $\beta - 1$ values. Therefore $2(\beta-1)\beta^{t-1}$ different numbers can be represented (the 2 accounts for the positive and negative sign). On the other hand, the exponent can assume $U - L + 1$ values. Thus, the set $\mathbb{F}(\beta, t, L, U)$ contains $2(\beta - 1)\beta^{t-1}(U - L + 1)$ different elements.

Solution 1.3 Use the instruction U=2*eye(10)-3*diag(ones(8,1),2) (respectively, L=2*eye(10)-3*diag(ones(8,1),-2)).

Solution 1.4 We can interchange the third and seventh rows of the previous matrix using the instructions: r=[1:10]; r(3)=7; r(7)=3; L(r,:). Notice that the character : in L(r,:) ensures that all columns of L are spanned in the usual increasing order (from the first to the last). To interchange the fourth column with the eighth column we can write c=[1:10]; c(8)=4; c(4)=8; L(:,c). Similar instructions can be used for the upper triangular matrix.

Solution 1.5 We can define the matrix A = [v1;v2;v3;v4] where v1, v2, v3 and v4 are the 4 given row vectors. They are linearly independent iff the determinant of A is different from 0, which is not true in our case.

Solution 1.6 The two given functions f and g have the symbolic expression:

```
>> syms x
>> f=sqrt(x^2+1); pretty(f)
```

$$(x^2+1)^{1/2}$$

```
>> g=sin(x^3)+cosh(x); pretty(g)
```

$$\sin(x^3) + \cosh(x)$$

syms   The command syms x allows one to use x as a symbolic variable. The com-
pretty mand pretty(f) prints the symbolic expression f in a format that resembles
type-set mathematics. At this stage, the symbolic expression of the first and
second derivatives and the integral of f can be obtained with the following
instructions:

```
>> diff(f,x)
ans =
1/(x^2+1)^(1/2)*x
>> diff(f,x,2)
ans =
-1/(x^2+1)^(3/2)*x^2+1/(x^2+1)^(1/2)
>> int(f,x)
ans =
1/2*x*(x^2+1)^(1/2)+1/2*asinh(x)
```

Similar instructions can be used for the function g.

Solution 1.7 The accuracy of the computed roots downgrades as the polynomial degree increases. This experiment reveals that the accurate computation
of the roots of a polynomial of high degree can be troublesome.

Solution 1.8 Here is a possible program to compute the sequence:

```
function I=sequence(n)
I = zeros(n+2,1); I(1) = (exp(1)-1)/exp(1);
for i = 0:n, I(i+2) = 1 - (i+1)*I(i+1); end
```

The sequence computed from this program doesn't tend to zero (as **n** increases), but it diverges with alternating sign.

Solution 1.9 The anomalous behavior of the computed sequence is due to
the propagation of roundoff errors from the innermost operation. In particular,
when $4^{1-n}z_n^2$ is less than ϵ_M, the elements of the sequence are equal to 0. This
happens from $n = 27$.

Solution 1.10 The proposed method is a special instance of a Monte Carlo
method and is implemented by the following program:

```
function mypi=pimontecarlo(n)
x = rand(n,1); y = rand(n,1);
z = x.^2+y.^2;
v = (z <= 1);
m=sum(v); mypi=4*m/n;
```

The command **rand** generates a sequence of pseudo-random numbers. The instruction v = (z <= 1) is a shortand version of the following procedure: we check whether z(k) <= 1 for any component of the vector z. If the inequality is satisfied for the k-th component of z (that is, the point (x(k),y(k)) belongs to the interior of the unit circle) v(k) is set equal to 1, and to 0 otherwise. The command **sum**(v) computes the sum of all components of v, that is, the number of points falling in the interior of the unit circle.

`sum`

By launching the program as mypi=pimontecarlo(n) for different values of n, when n increases, the approximation mypi of π becomes more accurate. For instance, for n=1000 we obtain mypi=3.1120, whilst for n=300000 we have mypi=3.1406.

Solution 1.11 The binomial coefficient can be computed by the following program (see also the MATLAB function nchoosek):

`nchoosek`

```
function bc=bincoeff(n,k)
k = fix(k); n = fix(n);
if k > n, disp('k must be between  0 and n'); break; end
if k > n/2, k = n-k; end
if k <= 1,  bc = n^k; else
  num = (n-k+1):n; den = 1:k; el = num./den; bc = prod(el);
end
```

The command **fix**(k) rounds k to the nearest integer smaller than k. The command **disp**(string) displays the string, without printing its name. In general, the command **break** terminates the execution of **for** and **while** loops. If **break** is executed in an **if**, it terminates the statement at that point. Finally, **prod**(el) computes the product of all elements of the vector el.

`fix`

`disp`

`break`

`prod`

9.2 Chapter 2

Solution 2.1 The command **fplot** allows us to study the graph of the given function f for various values of γ. For $\gamma = 1$, the corresponding function does not have real zeros. For $\gamma = 2$, there is only one zero, $\alpha = 0$, with multiplicity equal to 4 (that is, $f(\alpha) = f'(\alpha) = f''(\alpha) = f'''(\alpha) = 0$, while $f^{(4)}(\alpha) \neq 0$). Finally, for $\gamma = 3$, f has two distinct zeros, one in the interval $(-3, -1)$ and the other one in $(1, 3)$. In the case $\gamma = 2$, the bisection method cannot be used since it is impossible to find an interval (a, b) in which $f(a)f(b) < 0$. For $\gamma = 3$, starting from the interval $[a, b] = [-3, -1]$, the bisection method (Program 1) converges in 34 iterations to the value $\alpha = -1.85792082914850$ (with $f(\alpha) \simeq -3.6 \cdot 10^{-12}$), using the following instructions:

```
>> f=inline('cosh(x)+cos(x)-3'); a=-3; b=-1; tol=1.e-10; nmax=200;
>> [zero,res,niter]=bisection(f,a,b,tol,nmax)
zero =
   -1.8579
```

```
res =
    -3.6872e-12
niter =
    34
```

Similarly, for $\gamma = 4$ the bisection method converges after 34 iterations to the value $\alpha = -2.20562163688010$, with $f(\alpha) \simeq -1.8 \cdot 10^{-10}$.

Solution 2.2 We have to compute the zeros of the function $f(V) = pV + aN^2/V - abN^3/V^2 - pNb - kNT$. Plotting the graph of f, we see that this function has just a simple zero in the interval $(0.01, 0.06)$ with $f(0.01) < 0$ and $f(0.06) > 0$. We can compute this zero using the bisection method as follows:

```
>> f=inline('35000000*V+401000/V-17122.7/V^2-1494500','V');
>> [zero,res,niter]=bisection(f,0.01,0.06,1.e-12,100)
zero =
    0.0427
res =
    -6.3814e-05
niter =
    35
```

Solution 2.3 The unknown value of ω is the zero of the function $f(\omega) = s(1, \omega) - 1 = 9.8/2[\sinh(\omega) - \sin(\omega)]/\omega^2 - 1$. From the graph of f we conclude that f has a unique real zero in the interval $(0.5, 1)$. Starting from this interval, the bisection method computes the value $\omega = 0.61214447021484$ with the desired tolerance in 15 iterations as follows:

```
>> f=inline('9.8/2*(sinh (omega)- sin(omega))./omega.^2 -1','omega');
>> [zero,res,niter]=bisection(f,0.5,1,1.e-05,100)
zero =
    6.1214e-01
res =
    3.1051e-06
niter =
    15
```

The Newton method with initial value 0.5 would require only 3 iterations to provide the same significant digits, however with a much smaller residual:

```
>> fx=inline('(49/10*cosh(omega)-49/10*cos(omega))./omega.^2-2*...
             (49/10*sinh(omega)-49/10*sin(omega))./omega.^3','omega')
>> [zero,res,niter]=newton(f,fx,0.5,1.e-05,100)
zero =
    6.1214e-01
res =
    4.4409e-16
niter =
    3
```

Solution 2.4 The inequality (2.3) can be derived by observing that $|e^{(k)}| < |I^{(k)}|/2$ with $|I^{(k)}| < \frac{1}{2}|I^{(k-1)}| < 2^{-k}(b-a)$. Consequently, the error at the iteration k_{min} is less than ε if k_{min} is such that $2^{-k_{min}-1}(b-a) < \varepsilon$, that is, $2^{-k_{min}-1} < \varepsilon/(b-a)$, which proves (2.3).

Solution 2.5 The first formula is less sensitive to the roundoff error.

Solution 2.6 In Solution 2.1 we have analyzed the zeros of the given function with respect to different values of γ. Let us consider the case when $\gamma = 2$. Starting from the initial guess $x^{(0)} = 1$, the Newton method (Program 2) converges to the value $\bar{\alpha} = 1.0113e - 04$ in 31 iterations with `tol=1.e-10` while the exact zero of f is equal to 0. This big discrepancy is due to the fact that f is almost a constant in a neighborhood of its zero. Actually, the corresponding residual computed by MATLAB is 0. Let us set now $\gamma = 3$. The Newton method with `tol=1.e-16` converges to the value 1.85792082915020 in 10 iterations, while for $\gamma = 4$ it converges to 2.20562163692958 in 12 iterations (in both cases the residuals are zero in MATLAB).

Solution 2.7 The square and the cube roots of a number a are the solutions of the equations $x^2 = a$ and $x^3 = a$, respectively. Thus, the corresponding algorithms are: for a given $x^{(0)}$ compute

$$x^{(k+1)} = \frac{1}{2}\left(x^{(k)} + \frac{a}{x^{(k)}}\right), \quad k \geq 0 \qquad \text{for the square root,}$$

$$x^{(k+1)} = \frac{1}{3}\left(2x^{(k)} + \frac{a}{(x^{(k)})^2}\right), \quad k \geq 0 \quad \text{for the cube root.}$$

Solution 2.8 Setting $\delta x^{(k)} = x^{(k)} - \alpha$, from the Taylor expansion of f we find:

$$0 = f(\alpha) = f(x^{(k)}) + \delta x^{(k)} f'(x^{(k)}) + \frac{1}{2}(\delta x^{(k)})^2 f''(x^{(k)}) + \mathcal{O}((\delta x^{(k)})^3). \quad (9.1)$$

The Newton method yields

$$\delta x^{(k+1)} = \delta x^{(k)} - f(x^{(k)})/f'(x^{(k)}). \qquad (9.2)$$

Combining (9.1) with (9.2), we have

$$\delta x^{(k+1)} = \frac{1}{2}(\delta x^{(k)})^2 \frac{f''(x^{(k)})}{f'(x^{(k)})} + \mathcal{O}((\delta x^{(k)})^3).$$

After division by $(\delta x^{(k)})^2$ and letting $k \to \infty$ we prove the convergence result.

Solution 2.9 For certain values of β the equation (2.2) can have two roots that correspond to different configurations of the rods system. The two initial values that are suggested have been chosen conveniently to allow the Newton

method to converge toward one or the other root, respectively. We solve the
problem corresponding to the following values of β: $\beta = 2k\pi/300$ with $k =$
$0, \ldots, 100$. We use the following instructions to obtain the solution of the
problem (shown in Figure 9.1):

```
>> a = [10 13 8 10];
>> f = inline('a(1)/a(2)*cos(beta)-a(1)/a(4)*cos(alpha)-cos(beta-alpha)+...
   (a(1)^2+a(2)^2-a(3)^2+a(4)^2)/(2*a(2)*a(4))','alpha','beta','a');
>> df = inline('a(1)/a(2)*sin(alpha)-sin(beta-alpha)','alpha','beta','a');
>> tol = 1.e-05; x0 = -0.1; x1 = 2*pi/3; nmax = 200;
>> alpha1 = []; alpha2 = []; niter1 = []; niter2 = [];
>> vbeta = linspace(0,2*pi/3,100);
>> for beta = vbeta
   [zero,res,niter]=newton(f,df,x0,tol,nmax,beta,a);
   alpha1 = [alpha1,zero];   niter1 = [niter1,niter];
   [zero,res,niter]=newton(f,df,x1,tol,nmax,beta,a);
   alpha2 = [alpha2,zero];   niter2 = [niter2,niter];
   end
```

The components of the vectors `alpha1` and `alpha2` are the angles computed for
different values of β, while the components of the vectors `niter1` and `niter2`
are the number of Newton iterations (5 or 6) necessary to compute the zeros
with the requested tolerance.

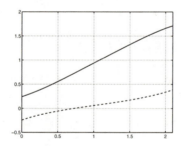

Fig. 9.1. The two curves represent the two possible configurations which
correspond to every value $\beta \in [0, 2\pi/3]$

Solution 2.10 From an inspection of its graph we see that f has two positive
real zeros ($\alpha_2 \simeq 1.5$ and $\alpha_3 \simeq 2.5$) and one negative ($\alpha_1 \simeq -0.5$). The Newton
method converges in 4 iterations (having set $x^{(0)} = -0.5$ and `tol = 1.e-10`)
to the value α_1:

```
>> f=inline('exp(x)-2*x^2'); df=inline('exp(x)-4*x'); x0=-0.5; tol=1.e-10;
>> nmax=100; format long; [zero,res,niter]=newton(f,df,x0,tol,nmax)
zero =
  -0.53983527690282
res =
```

```
        0
niter =
        4
```

The given function has a maximum at $\bar{x} \simeq 0.3574$ (which can be obtained by applying the Newton method to the function f'): for $x^{(0)} < \bar{x}$ the method converges to the negative zero. If $x^{(0)} = \bar{x}$ the Newton method cannot be applied since $f'(\bar{x}) = 0$.

Solution 2.11 Let us set $x^{(0)} = 0$ and tol$= 10^{-17}$. The Newton method converges in 32 iterations to the value 0.64118239763649, which we identify with the exact zero α. We can observe that the (approximate) errors $x^{(k)} - \alpha$, for $k = 1, 2, \ldots, 31$, decrease only linearly when k increases. This behavior is due to the fact that α has multiplicity greater than 1 (see Figure 9.2). To recover a second-order method we can use the modified Newton method.

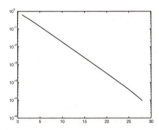

Fig. 9.2. Error vs iteration number of the Newton method for the computation of the zero of the function $f(x) = x^3 - 3x^2 2^{-x} + 3x4^{-x} - 8^{-x}$

Solution 2.12 We should compute the zero of the function $f(x) = \sin(x) - \sqrt{2gh/v_0^2}$. From an inspection of its graph, we can conclude that f has one zero in the interval $(0, \pi/2)$. The Newton method with $x^{(0)} = \pi/4$ and tol$= 10^{-10}$ converges in 5 iterations to the value 0.45862863227859.

Solution 2.13 Using the data given in the exercise, the solution can be obtained with the following instructions:

```
>> f=inline('M-v*(1+I).*((1+I).^n - 1)./I','I','M','v','n');
>> df=inline('-v*((1+I).^n-1)./I-n*v*(1+I).^n./I+v*(1+I).*((1+I).^n-1)./I.^2',...
   'I','M','v','n');
>> [zero,res,niter]=bisection(f,0.01,0.1,1.e-12,4,6000,1000,5);
>> [zero,res,niter]=newton(f,df,zero,1.e-12,100,6000,1000,5);
```

The Newton method converges to the desired result in 3 iterations.

Solution 2.14 By a graphical study, we see that (2.15) is satisfied for a value of α in $(\pi/6, \pi/4)$. The Newton method provides the approximate value

0.59627992746547 in 5 iterations, starting from $x^{(0)} = \pi/4$. We deduce that the maximum length of a rod that can pass in the corridor is $L = 30.84$.

Solution 2.15 If α is a zero of f with multiplicity m, then there exists a function h such that $h(\alpha) \neq 0$ and $f(x) = h(x)(x - \alpha)^m$. The desired result can be obtained by writing both f and its first derivative in the Newton method in terms of h. We obtain for the iteration function the following expression

$$\phi_N(x) = x - \frac{h(x)(x-\alpha)^m}{h'(x)(x-\alpha)^m + mh(x)(x-\alpha)^{m-1}} = x - \frac{h(x)(x-\alpha)}{h'(x)(x-\alpha) + mh(x)}$$

By computing its first derivative, we have:

$$\phi_N'(x) = 1 - \frac{[h'(x)(x-\alpha) + h(x)]D(x) - h(x)(x-\alpha)D'(x)}{[D(x)]^2},$$

where $D(x) = h'(x)(x-\alpha) + mh(x)$. Therefore $\phi_N'(\alpha) = 1 - 1/m$ and thus the Newton method converges quadratically only if $m = 1$.

Solution 2.16 Let us inspect the graph of f by using the following commands:

```
>> f= 'x^3+4*x^2-10'; fplot(f,[-10,10]); grid on;
>> fplot(f,[-5,5]); grid on;
>> fplot(f,[0,5]); grid on
```

We can see that f has only one real zero, equal approximately to 1.36 (see Figure 9.3). The iteration function and its derivative are:

$$\phi(x) = \frac{2x^3 + 4x^2 + 10}{3x^2 + 8x},$$

$$\phi'(x) = \frac{(6x^2 + 8x)(3x^2 + 8x) - (6x + 8)(2x^3 + 4x^2 + 10)}{(3x^2 + 8x)^2},$$

and $\phi(\alpha) = \alpha$. We easily deduce that $\phi'(\alpha) = 0$ by noting that $\phi'(x) = (6x + 8)f(x)/(3x^2 + 8x)^2$. Consequently, the proposed method converges (at least) quadratically.

Solution 2.17 The proposed method is convergent at least with order 2 since $\phi'(\alpha) = 0$.

Solution 2.18 By keeping the remaining parameters unchanged, the method converges after only 3 iterations to the value 0.64118573649623 which differs by less than 10^{-9} from the result previously computed. However, the behavior of the function, which is quite flat near to $x = 0$, suggests that the result computed previously could be more accurate. In Figure 9.4 we show the graph of f in $(0.5, 0.7)$, obtained with the following instructions:

```
>> f='x^3-3*x^2*2^(-x) + 3*x*4^(-x) - 8^(-x)';
>> fplot(f,[0.5 0.7]); grid on
```

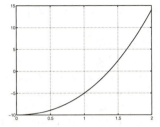

Fig. 9.3. Graph of $f(x) = x^3 + 4x^2 - 10$ for $x \in [0, 2]$

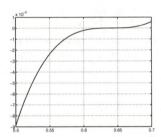

Fig. 9.4. Graph of $f(x) = x^3 - 3x^2 2^{-x} + 3x 4^{-x} - 8^{-x}$ for $x \in [0.5, 0.7]$

9.3 Chapter 3

Solution 3.1 Since $x \in (x_0, x_n)$, there exists an interval $I_i = (x_{i-1}, x_i)$ such that $x \in I_i$. We can easily see that $\max_{x \in I_i} |(x - x_{i-1})(x - x_i)| = h^2/4$. If we bound $|x - x_{i+1}|$ above by $2h$, $|x - x_{i-2}|$ by $3h$ and so on, we obtain the inequality (3.6).

Solution 3.2 In all cases we have $n = 4$ and thus we should estimate the fifth derivative of each function in the given interval. We find: $\max_{x \in [-1,1]} |f_1^{(5)}| <$ 1.76, $\max_{x \in [-1,1]} |f_2^{(5)}| < 1.55$, $\max_{x \in [-\pi/2, \pi/2]} |f_3^{(5)}| < 1.42$. The corresponding errors are therefore bounded by 0.0028, 0.0024 and 0.0212, respectively.

Solution 3.3 Using the command `polyfit` we compute the interpolating polynomials of degree 3 in the two cases:

```
>> years=[1975 1980 1985 1990];
>> east=[70.2 70.2 70.3 71.2];
>> west=[72.8 74.2 75.2 76.4];
>> ceast=polyfit(years,eor,3);
>> cwest=polyfit(years,eoc,3);
>> esteast=polyval(ceast,[1970 1983 1988 1995])
esteast =
   69.6000   70.2032   70.6992   73.6000
```

```
>> estwest=polyval(cwest,[1970 1983 1988 1995])
estwest =
   70.4000   74.8096   75.8576   78.4000
```

Thus, for Western Europe the life expectation in the year 1970 is equal to 70.4 years (`estwest(1)`), with a discrepancy of 1.4 years from the real value. The symmetry of the graph of the interpolating polynomial suggests that the estimation for the life expectation of 78.4 years for the year 1995, can be overestimated by the same quantity (in fact, the real life expectation is equal to 77.5 years). A different conclusion holds concerning Eastern Europe. Indeed, in that case the estimation for 1970 coincides exactly with the real value, while the estimation for 1995 is largely overestimated (73.6 years instead of 71.2).

Solution 3.4 We chose the month as time-unit. The initial time $t_0 = 1$ corresponds to November 1987, while $t_7 = 157$ to November 2000. With the following instructions we compute the coefficients of the polynomial interpolating the given prices:

```
>> time = [1 14 37 63 87 99 109 157];
>> price = [4.5 5 6 6.5 7 7.5 8 8];
>> [c] = polyfit(time,price,7);
```

Setting `[price2002]= polyval(c,180)` we find that the estimated price of the magazine in November 2002 is approximately 10.8 euros.

Solution 3.5 The interpolatory cubic spline, computed by the command `spline` in this special case, coincides with the interpolating polynomial. This wouldn't be true for the natural interpolating cubic spline.

Solution 3.6 We use the following instructions:

```
>> T = [4:4:20];
>> rho=[1000.7794,1000.6427,1000.2805,999.7165,998.9700];
>> Tnew = [6:4:18]; format long e;
>> rhonew = spline(T,rho,Tnew)
rhonew =
  Columns 1 through 3
    1.000740787500000e+03   1.000488237500000e+03   1.000022450000000e+03
  Column 4
    9.993649250000000e+02
```

The comparison with the further measures shows that the approximation is extremely accurate. Note that the state equation for the sea-water (UNESCO, 1980) assumes a fourth-order dependence of the density on the temperature. However, the coefficient of the fourth power of T is of order of 10^{-9}.

Solution 3.7 To compute the interpolatory cubic spline we use the following instructions:

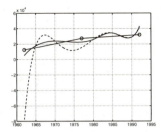

Fig. 9.5. The cubic spline (continuous line) and interpolating polynomial (dotted line) for the data of Exercise 3.7. The circles denote the values used in the interpolation

```
>> year=[1965 1970 1980 1985 1990 1991];
>> production=[17769 24001 25961 34336 29036 33417];
>> years=[1962 1977 1992];
>> estimateprod=spline(year,production,years);
```

We obtain the following values: 5146.1×10^5 Kg in 1962; 22641.7×10^5 Kg in 1977; 41894.4×10^5 Kg in 1992. The comparison with the real data (12380, 27403 and 32059×10^5 Kg, respectively) shows that the values predicted by the spline are inaccurate outside the interpolation interval (see Figure 9.5). Nonetheless, the control exerted by the spline at the end-points of the interpolation interval allows us to obtain reasonable values also for 1965. On the contrary, the interpolating polynomial introduces large oscillations near this end-point and underestimates the production of as many as -77685×10^5 Kg for 1962.

Solution 3.8 The interpolating polynomial p and the spline s3 at 21 equispaced nodes in $[-1, 1]$, can be evaluated in 201 equispaced nodes by the following instructions:

```
>> pert = 1.e-04;
>> x=[-1:2/20:1]; y=sin(2*pi*x)+(-1).^[1:21]*pert; z=[-1:0.01:1];
>> c=polyfit(x,y,20); p=polyval(c,z); s3=spline(x,y,z);
```

When we use the unperturbed data (`pert=0`) the graphs of both p and s3 are indistinguishable from that of the given function. The situation changes dramatically when the perturbed data are used (`pert=1.e-04`). In particular, the interpolating polynomial shows strong oscillations at the end-points of the interval, whereas the spline remains practically unchanged (see Figure 9.6). This example shows that approximation by splines is in general more stable with respect to perturbation errors.

Solution 3.9 If $n = m$, setting $\tilde{f} = \Pi_n f$ we find that the first member of (3.18) is null. Thus in this case $\Pi_n f$ is the unique solution of the least-square problem.

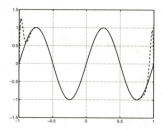

Fig. 9.6. The interpolating polynomial (dotted line) and the interpolatory cubic spline (continuous line) corresponding to the perturbed data. Note the severe oscillations of the interpolating polynomial near the end-points of the interval

Solution 3.10 The coefficients (obtained by the command `polyfit`) of the requested polynomials are (only the first 4 significant digits are shown):

$K = 0.67$, $a_4 = 6.301\ 10^{-8}$, $a_3 = -8.320\ 10^{-8}$, $a_2 = -2.850\ 10^{-4}$, $a_1 = 9.718\ 10^{-4}$, $a_0 = -3.032$;

$K = 1.5$, $a_4 = -4.225\ 10^{-8}$, $a_3 = -2.066\ 10^{-6}$, $a_2 = 3.444\ 10^{-4}$, $a_1 = 3.364 10^{-3}$, $a_0 = 3.364$;

$K = 2$, $a_4 = -1.012\ 10^{-7}$, $a_3 = -1.431\ 10^{-7}$, $a_2 = 6.988\ 10^{-4}$, $a_1 = -1.060\ 10^{-4}$, $a_0 = 4.927$;

$K = 3$, $a_4 = -2.323\ 10^{-7}$, $a_3 = 7.980\ 10^{-7}$, $a_2 = 1.420\ 10^{-3}$, $a_1 = -2.605\ 10^{-3}$, $a_0 = 7.315$.

In Figure 9.7 we show the graph of the polynomial computed using the data in the first column of Table 3.1.

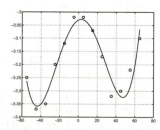

Fig. 9.7. Least-square polynomial of degree 4 (continuous line) compared with the data in the first column of Table 3.1

Solution 3.11 By repeating the first 3 instructions reported in Solution 3.7 and using the command `polyfit`, we find the following values (in 10^5 Kg): 15280.12 in 1962; 27407.10 in 1977; 32019.01 in 1992, which represent good approximations to the real ones (12380, 27403 and 32059, respectively).

Solution 3.12 We can rewrite the coefficients of the system (3.20) in terms of mean and variance by noting that the variance can be expressed as $v = \frac{1}{n+1}\sum_{i=0}^{n} x_i^2 - M^2$.

Solution 3.13 The desired property is deduced from the first equation of the system that provides the coefficients of the least-squares straight line.

9.4 Chapter 4

Solution 4.1 Using the following third-order Taylor expansions of f at the point x_0, we obtain

$$4f(x_1) = 4f(x_0) + 4hf'(x_0) + 2h^2 f''(x_0) + \frac{2}{3}h^3 f'''(\xi),$$

$$-f(x_2) = -f(x_0) - 2hf'(x_0) - 2h^2 f''(x_0) - \frac{4}{3}h^3 f'''(\eta).$$

with $\xi \in (x_0, x_1)$ and $\eta \in (x_0, x_2)$ as two suitable points. Summing this two expressions yields

$$-3f(x_0) + 4f(x_1) - f(x_2) = 2hf'(x_0) + \frac{2h^3}{3}(f'''(\xi) - 2f'''(\eta)).$$

Dividing by $2h$ and using the mean value theorem, we can conclude the exercise with $\xi_0 \in (x_0, x_2)$. A similar procedure can be used for the formula at x_n.

Solution 4.2 Taylor expansions yield

$$f(\bar{x} + h) = f(\bar{x}) + hf'(\bar{x}) + \frac{h^2}{2}f''(\bar{x}) + \frac{h^3}{6}f'''(\xi),$$

$$f(\bar{x} - h) = f(\bar{x}) - hf'(\bar{x}) + \frac{h^2}{2}f''(\bar{x}) - \frac{h^3}{6}f'''(\eta),$$

where ξ and η are suitable points. Subtracting these two expressions and dividing by $2h$ we obtain the result (4.9).

Solution 4.3 Assuming that $f \in C^4$ and proceeding as in Solution 4.2 we obtain the following errors (for suitable points ξ_1, ξ_2 and ξ_3):

$$a. \ -\frac{1}{4}f^{(4)}(\xi_1)h^3, \quad b. \ -\frac{1}{12}f^{(4)}(\xi_2)h^3, \quad c. \ \frac{1}{30}f^{(4)}(\xi_3)h^4.$$

Solution 4.4 Using the approximation (4.8), we obtain the following values:

t (months)	0	0.5	1	1.5	2	2.5	3
δn	--	78.04	44.95	19.28	7.02	2.39	--
n'	--	77.78	39.09	15.19	5.29	1.77	--

By comparison with the exact values of $n'(t)$ we can conclude that the computed values are sufficiently accurate.

Solution 4.5 The quadrature error can be bounded by

$$(b - a)^3/(24M^2) \max_{x \in [a,b]} |f''(x)|,$$

where $[a, b]$ is the integration interval and M the (unknown) number of subintervals.

The function f_1 is infinitely differentiable. From the graph of f_1'' we infer that $|f_1''(x)| \leq 2$ in the integration interval. Thus the integration error for f_1 is less than 10^{-4} provided that $5^3/(24M^2)2 < 10^{-4}$, that is $M > 322$.

Also the function f_2 is differentiable to any order. Since $\max_{x \in [0,\pi]} |f_2''(x)| = \sqrt{2}e^{3/4\pi}$, the integration error is less than 10^{-4} provided that $M > 439$. These inequalities actually provide an over estimation of the integration errors. Indeed, the (effective) minimum number of intervals which ensures that the error is below the fixed tolerance of 10^{-4} is much lower than that predicted by our result (for instance, for the function f_1 this number is 51). Finally, we note that since f_3 is not differentiable in the integration interval, our theoretical error estimate doesn't hold.

Solution 4.6 On each interval I_k, $k = 1, \ldots, M$, the error committed is equal to $H^3/24 f''(\xi_k)$ with $\xi_k \in (x_{k-1}, x_k)$ and hence the global error will be $H^3/24 \sum_{k=1}^{M} f''(\xi_k)$. Since f'' is a continuous function in (a, b) there exists a point $\xi \in (a, b)$ such that $f''(\xi) = \frac{1}{M} \sum_{k=1}^{M} f''(\xi_k)$. Using this result and the fact that $MH = b - a$, we derive equation (4.13).

Solution 4.7 This effect is due to the accumulation of local errors on each sub-interval.

Solution 4.8 By construction, the mid-point formula integrates exactly the constants. To verify that the linear polynomials also are exactly integrated, it is sufficient to verify that $I(x) = I_{PM}(x)$. As a matter of fact we have

$$I(x) = \int_a^b x \, dx = \frac{b^2 - a^2}{2}, \quad I_{PM}(x) = (b - a)\frac{b + a}{2}.$$

Solution 4.9 For the function f_1 we find $M = 71$ if we use the trapezoidal formula and only $M = 7$ for the Gauss formula. Indeed, the computational advantage of this latter formula is evident.

Solution 4.10 Equation (4.16) states that the quadrature error for the composite trapezoidal formula with $H = H_1$ is equal to CH_1^2, with $C = -\dfrac{b - a}{12} f''(\xi)$. If f'' does not vary "too much", we can assume that also the error with $H = H_2$ behaves like CH_2^2. Then, by equating the two expressions

$$I(f) \simeq I_1 + CH_1^2, \quad I(f) \simeq I_2 + CH_2^2, \tag{9.3}$$

we obtain $C = (I_1 - I_2)/(H_2^2 - H_1^2)$. Using this value in one of the expressions (9.3), we obtain equation (4.29), that is, a better approximation than is produced by I_1 or I_2.

Solution 4.11 We seek the maximum positive integer p such that $I_{approx}(x^p) = I(x^p)$. For $p = 0, 1, 2, 3$ we find the following non-linear system with 4 equations in the 4 unknowns α, β, $\bar{x}$ and $\bar{z}$:

$$p = 0 \quad \rightarrow \quad \alpha + \beta = b - a,$$
$$p = 1 \quad \rightarrow \quad \alpha\bar{x} + \beta\bar{z} = \frac{b^2 - a^2}{2},$$
$$p = 2 \quad \rightarrow \quad \alpha\bar{x}^2 + \beta\bar{z}^2 = \frac{b^3 - a^3}{3},$$
$$p = 3 \quad \rightarrow \quad \alpha\bar{x}^3 + \beta\bar{z}^3 = \frac{b^4 - a^4}{4}.$$

From the first two equations we can eliminate α and $\bar{z}$ and reduce the system to a new one in the unknowns β and $\bar{x}$. In particular, we find a second-order equation in β from which we can compute β as a function of $\bar{x}$. Finally, the non-linear equation in $\bar{x}$ can be solved by the Newton method, yielding two values of $\bar{x}$ that are the abscissae of the Gauss quadrature points.

Solution 4.12 Since

$$f_1^{(4)}(x) = \frac{24}{(1 + (x - \pi)^2)^5(2x - 2\pi)^4} - \frac{72}{(1 + (x - \pi)^2)^4(2x - 2\pi)^2}$$
$$+ \frac{24}{(1 + (x - \pi)^2)^3},$$
$$f_2^{(4)}(x) = -4e^x \cos(x),$$

we find that the maximum of $|f_1^{(4)}(x)|$ is bounded by $M_1 \simeq 25$, while that of $|f_2^{(4)}(x)|$ by $M_2 \simeq 93$. Consequently, from (4.22) we obtain $H < 0.21$ in the first case and $H < 0.16$ in the second case.

Solution 4.13 Using the command int('exp(-x^2/2)',0,2) we obtain for the integral at hand the value 1.19628801332261.
   The Gauss formula (4.20) applied to the same interval would provide the value 1.20278027622354 (with an absolute error equal to 6.4923e-03), while the Simpson formula gives 1.18715264069572 with a slightly larger error (equal to 9.1354e-03).

Solution 4.14 We note that $I_k > 0 \ \forall k$, since the integrand is non-negative. Therefore, we expect that all the values produced by the recursive formula should be non-negative. Unfortunately, the recursive formula is unstable to the propagation of roundoff errors and produces negative elements:

```
>> l(1)=1/exp(1); for k=2:20, l(k)=1-k*l(k-1); end
>> l(20)
 -30.1924
```

Using the composite Simpson formula, with $H < 0.25$, we can compute the integral with the desired accuracy.

Solution 4.15 For the Simpson formula we obtain

$$I_1 = 1.19616568040561, \quad I_2 = 1.19628173356793, \quad \Rightarrow \quad I_R = 1.19628947044542,$$

with an absolute error in I_R equal to -1.4571e-06 (we earn two orders of magnitude with respect to I_1 and a factor $1/4$ with respect to I_2). Using the Gauss formula we obtain (the errors are reported between brackets):

$$\begin{aligned} I_1 &= 1.19637085545393 & (-8.2842e-05), \\ I_2 &= 1.19629221796844 & (-4.2046e-06), \\ I_R &= 1.19628697546941 & (1.0379e-06). \end{aligned}$$

The advantage of using the Richardson extrapolation method is evident.

Solution 4.16 We must compute by the Simpson formula the values $j(r) = \sigma/(\varepsilon_0 r^2) \int_0^r f(\xi)d\xi$ with $r = k/10$, for $k = 1, \ldots, 10$ and $f(\xi) = e^\xi \xi^2$.

In order to estimate the integration error we need the fourth derivative $f^{(4)}(\xi) = e^\xi(\xi^2 + 8\xi + 12)$. The maximum of $f^{(4)}$ in the integration interval $(0, r)$ is attained at $\xi = r$ since $f^{(4)}$ is monotonically increasing. Then we obtain the following values:

```
>> r=[0.1:0.1:1];
>> maxf4=exp(r).*(r.^2+8*r+12);
maxf4 =
  Columns 1 through 7
   14.1572   16.6599   19.5595   22.9144   26.7917   31.2676   36.4288
  Columns 8 through 10
   42.3743   49.2167   57.0839
```

For a given r the error is below 10^{-10} provided that $H_r^4 < 10^{-10}2880/(rf^{(4)}(r))$. For $r = k/10$ with $k = 1, \ldots, 10$ by the following instructions we can compute the minimum numbers of subintervals which ensure that the previous inequalities are satisfied. The components of the vector M contain these numbers:

```
>> x=[0.1:0.1:1]; f4=exp(x).*(x.^2+8*x+12);
>> H=(10^(-10)*2880./(x.*f4)).^(1/4); M=fix(x./H)
M =
    4    11    20    30    41    53    67    83   100   118
```

Therefore, the values of $j(r)$ are:

```
>> sigma=0.36; epsilon0 = 8.859e-12;
   f = inline('exp(x).*x.^2');
```

```
for k = 1:10
   r = k/10;
   j(k)=simpsonc(0,r,M(k),f);
   j(k) = j(k)*sigma/r*epsilon0;
end
```

Solution 4.17 We compute $E(213)$ using the Simpson composite formula by increasing the number of intervals until the difference between two consecutive approximations (divided by the last computed value) is less than 10^{-11}:

```
>> f=inline('2.39e-11./((x.^5).*(exp(1.432./(T*x))-1))','x','T');
>> a=3.e-04; b=14.e-04; T=213;
>> i=1; err = 1; Iold = 0; while err >= 1.e-11
I=simpsonc(a,b,i,f,T);
err = abs(I-Iold)/abs(I);
Iold=I;
i=i+1;
end
```

The procedure returns the value $i = 59$. Therefore, using 58 equispaced intervals we can compute the integral $E(213)$ with ten exact significant digits. The same result could be obtained by the Gauss formula using 53 intervals. Note that as many as 1609 intervals would be nedeed if using the composite trapezoidal formula.

Solution 4.18 On the whole interval the given function is not regular enough to allow the application of the theoretical convergence result (4.22). One possibility is to decompose the integral into the sum of two intervals, $(0, 0.5)$ and $(0.5, 1)$, in which the function is regular (it is actually a polynomial of degree 3). In particular, if we use the Simpson rule on each interval we can even integrate f exactly.

9.5 Chapter 5

Solution 5.1 The number r_k of algebraic operations (sums, subtractions and multiplications) required to compute a determinant of a matrix of order $k \geq 2$ with the Laplace rule (1.8), satisfies the following difference equation:

$$r_k - kr_{k-1} = 2k - 1,$$

with $r_1 = 0$. Multiplying both side of this equation by $1/k!$, we obtain

$$\frac{r_k}{k!} - \frac{r_{k-1}}{(k-1)!} = \frac{2k-1}{k!}.$$

Summing both sides from 2 to n gives the solution:

$$r_n = n! \sum_{k=2}^{n} \frac{2k-1}{k!} = n! \sum_{k=1}^{n-1} \frac{2k+1}{(k+1)!}, \qquad n \geq 1.$$

Solution 5.2 We use the following MATLAB commands to compute the determinants and the corresponding CPU-times:

```
>> t = [ ]; for i = 3:500
     A = magic(i); tt = cputime; d=det(A); t=[t, cputime-tt];
   end
```

The coefficients of the cubic least-square polynomial that approximate the data n=[3:500] and t are

```
>> format long; c=polyfit(n,t,3)
c =
   0.00000002102187  0.00000171915661  -0.00039318949610  0.01055682398911
```

The first coefficient (that multiplies n^3), is small, but not small enough with respect to the second one to be neglected. Indeed, if we compute the fourth degree least-square polynomial we obtain the following coefficients:

```
>> c=polyfit(i,t,4)
c =
   Columns 1 through 4
   -0.00000000000051  0.00000002153039  0.00000155418071  -0.00037453657810
   Column 5
    0.01006704351509
```

From this result, we can conclude that the computation of a determinant of a matrix of dimension n requires approximately n^3 ops.

Solution 5.3 We have: $\det A_1 = 1$, $\det A_2 = \varepsilon$, $\det A_3 = \det A = 2\varepsilon + 12$. Consequently, if $\varepsilon = 0$ the second principal submatrix is singular and the Proposition 5.1 cannot be applied. The matrix is singular if $\varepsilon = -6$. In this case the Gauss factorization yields

$$L = \begin{bmatrix} 1 & 0 & 0 \\ 2 & 1 & 0 \\ 3 & 1.25 & 1 \end{bmatrix}, \quad U = \begin{bmatrix} 1 & 7 & 3 \\ 0 & -12 & -4 \\ 0 & 0 & 0 \end{bmatrix}.$$

Note that U is singular (as we could have predicted since A is singular).

Solution 5.4 At step 1, $n-1$ divisions were used to calculate the l_{1k} entries for $i = 2, \ldots, n$. Then $(n-1)^2$ multiplications and $(n-1)^2$ additions were used to create the new entries $a_{ij}^{(2)}$, for $j = 2, \ldots, n$. At step 2, the numbers of divisions is $(n-2)$, while the numbers of multiplications and addictions will be $(n-2)^2$. At final step $n-1$ only 1 addiction, 1 multiplication and 1 division is required. Thus, using the identies

$$\sum_{s=1}^{q} s = \frac{q(q+1)}{2}, \quad \sum_{s=1}^{q} s^2 = \frac{q(q+1)(2q+1)}{6}, \quad q \geq 1,$$

we can conclude that to complete the Gaussian factorization $2(n - 1)n(n + 1)/3 + n(n - 1)$ *ops* are required. Neglecting the lower order terms, we can state that the Gaussian factorization process has a cost of $2n^3/3$ *ops*.

Solution 5.5 By definition, the inverse X of a matrix $A \in \mathbb{R}^{n \times n}$ satisfies $XA = AX = I$. Therefore, for $j = 1, \ldots, n$ the column vector $\mathbf{y}_j$ of X is the solution of the linear system $A\mathbf{y}_j = \mathbf{e}_j$, where $\mathbf{e}_j$ is the j-th vector of the canonical basis of $\mathbb{R}^n$ with all components equal to zero except the j-th that is equal to 1. After computing the LU factorization of A, the computation of the inverse of A requires the solution of n linear systems with the same matrix and different right-hand sides.

Solution 5.6 Using the Program 7 we compute the following factors:

$$
L = \begin{bmatrix} 1 & 0 & 0 \\ 2 & 1 & 0 \\ 3 & -3.38 \cdot 10^{15} & 1 \end{bmatrix}, \quad U = \begin{bmatrix} 1 & 1 & 3 \\ 0 & -8.88 \cdot 10^{-16} & 14 \\ 0 & 0 & 4.73 \cdot 10^{-16} \end{bmatrix}.
$$

If we compute their product we obtain the matrix

```
>> L*U
ans =
   1.0000   1.0000    3.0000
   2.0000   2.0000   20.0000
   3.0000   6.0000   -2.0000
```

which differs from A since the entry in position (3,3) is equal to -2 while in A it is equal to 4.

Solution 5.7 Usually, only the triangular (upper or lower) part of a symmetric matrix is stored. Therefore, any operation that does not respect the symmetry of the matrix is not optimal in view of the memory storage. This is the case when row pivoting is carried out. A possibility is to exchange simultaneously rows and columns having the same index, limiting therefore the choice of the pivot only to the diagonal elements. More generally, a pivoting strategy involving exchange of rows and columns is called *complete pivoting* (see, *e.g.*, [QSS00, Chap. 3]).

Solution 5.8 The L and U factors are:

$$
L = \begin{bmatrix} 1 & 0 & 0 \\ (\varepsilon - 2)/2 & 1 & 0 \\ 0 & -1/\varepsilon & 1 \end{bmatrix}, \quad U = \begin{bmatrix} 2 & -2 & 0 \\ 0 & \varepsilon & 0 \\ 0 & 0 & 3 \end{bmatrix}.
$$

When $\varepsilon \to 0$ $l_{32} \to \infty$. In spite of that, the solution of the system is accurate also when ε tends to zero as confirmed by the following instructions:

```
>> e=1; for k=1:10
b=[0; e; 2];
L=[1 0 0; (e-2)*0.5 1 0; 0 -1/e 1]; U=[2 -2 0; 0 e 0; 0 0 3];
y=L\b; x=U\y; err(k)=max(abs(x-ones(3,1))); e=e*0.1;
end
>> err
err =
      0    0    0    0    0    0    0    0    0    0
```

Solution 5.9 The computed solutions become less and less accurate when i increases. Indeed, the error norms are equal to $2.63 \cdot 10^{-14}$ for $i = 1$, to $9.89 \cdot 10^{-10}$ for $i = 2$ and to $2.10 \cdot 10^{-6}$ for $i = 3$. This can be explained by observing that the condition number of A_i increases as i increases. Indeed, using the command cond we find that the condition number of A_i is $\simeq 10^3$ for $i = 1$, $\simeq 10^7$ for $i = 2$ and $\simeq 10^{11}$ for $i = 3$.

Solution 5.10 If $(\lambda, \mathbf{v})$ are an eigenvalue-eigenvector pair of a matrix A, then λ^2 is an eigenvalue of A^2 with the same eigenvector. Indeed, from $A\mathbf{v} = \lambda\mathbf{v}$ follows $A^2\mathbf{v} = \lambda A\mathbf{v} = \lambda^2\mathbf{v}$. Consequently, if A is symmetric and positive definite $K(A^2) = (K(A))^2$.

Solution 5.11 The iteration matrix of the Jacobi method is:

$$
B_J = \begin{bmatrix} 0 & 0 & -\alpha^{-1} \\ 0 & 0 & 0 \\ -\alpha^{-1} & 0 & 0 \end{bmatrix}.
$$

Its eigenvalues are $\{0, \alpha^{-1}, -\alpha^{-1}\}$. Thus the method converges if $|\alpha| > 1$.
  The iteration matrix of the Gauss-Seidel method is

$$
B_{GS} = \begin{bmatrix} 0 & 0 & -\alpha^{-1} \\ 0 & 0 & 0 \\ 0 & 0 & \alpha^{-2} \end{bmatrix}
$$

with eigenvalues $\{0, 0, \alpha^{-2}\}$. Therefore, the method converges if $|\alpha| > 1$. In particular, since $\rho(B_{GS}) = [\rho(B_J)]^2$, the Gauss-Seidel converges more rapidly than the Jacobi method.

Solution 5.12 A sufficient condition for the convergence of the Jacobi and the Gauss-Seidel methods is that A is strictly diagonally dominant. The second row of A satisfies the condition of diagonal dominance provided that $|\beta| < 5$. Note that if we require directly that the spectral radii of the iteration matrices are less than 1 (which is a sufficient and necessary condition for convergence), we find the (less restrictive) limitation $|\beta| < 25$ for both methods.

Solution 5.13 The relaxation method in vector form is

$$
(I - \omega D^{-1}E)\mathbf{x}^{(k+1)} = [(1 - \omega)I + \omega D^{-1}F]\mathbf{x}^{(k)} + \omega D^{-1}\mathbf{b}
$$

where $A = D - E - F$, D being the diagonal of A, and E and F the lower (resp. upper) part of A. The corresponding iteration matrix is

$$B(\omega) = (I - \omega D^{-1}E)^{-1}[(1 - \omega)I + \omega D^{-1}F].$$

If we denote by λ_i the eigenvalues of $B(\omega)$, we obtain

$$\left| \prod_{i=1}^{n} \lambda_i \right| = \left| \det \left[(1 - \omega)I + \omega D^{-1}F \right] \right| = |1 - \omega|^n.$$

Therefore, at least one eigenvalue must satisfy the inequality $|\lambda_i| \geq |1 - \omega|$. Thus, a necessary condition to ensure convergence is that $|1 - \omega| < 1$, that is, $0 < \omega < 2$.

Solution 5.14 The given matrix is symmetric. To verify whether it is also definite positive, that is, $z^T A z > 0$ for all $z \neq 0$ of $\mathbb{R}^2$, we use the following instructions:

```
>> syms z1 z2 real
>> z=[z1;z2]; A=[3 2; 2 6];
>> pos=z'*A*z; simple(pos)
ans =
3*z1^2+4*z1*z2+6*z2^2
```

The command `syms z1 z2 real` is necessary to declare that the symbolic variables `z1` and `z2` are real numbers, while the command `simple(pos)` tries several algebraic simplifications of `pos` and returns the shortest. It is easy to see that the computed quantity is positive since it can be rewritten as `2*(z1+z2)^2 +z1^2+4*z2^2`. Thus, the given matrix is symmetric and positive definite, and the Gauss-Seidel method is convergent.

Solution 5.15 We find:

for the Jacobi method:
$$\begin{cases} x_1^{(1)} = \frac{1}{2}(1 - x_2^{(0)}), \\ x_2^{(1)} = -\frac{1}{3}(x_1^{(0)}); \end{cases} \Rightarrow \begin{cases} x_1^{(1)} = \frac{1}{4}, \\ x_2^{(1)} = -\frac{1}{3}; \end{cases}$$

for the Gauss-Seidel method:
$$\begin{cases} x_1^{(1)} = \frac{1}{2}(1 - x_2^{(0)}), \\ x_2^{(1)} = -\frac{1}{3}x_1^{(1)}, \end{cases} \Rightarrow \begin{cases} x_1^{(1)} = \frac{1}{4}, \\ x_2^{(1)} = -\frac{1}{12}. \end{cases}$$

For the gradient method, we first compute the initial residual

$$r^{(0)} = b - Ax^{(0)} = \begin{bmatrix} 1 \\ 0 \end{bmatrix} - \begin{bmatrix} 2 & 1 \\ 1 & 3 \end{bmatrix} x^{(0)} = \begin{bmatrix} -3/2 \\ -5/2 \end{bmatrix}.$$

Then, since

$$P^{-1} = \begin{bmatrix} 1/2 & 0 \\ 0 & 1/3 \end{bmatrix},$$

we have $\mathbf{z}^{(0)} = P^{-1}\mathbf{r}^{(0)} = (-3/4, -5/6)^T$. Therefore

$$\alpha_0 = \frac{(\mathbf{z}^{(0)})^T\mathbf{r}^{(0)}}{(\mathbf{z}^{(0)})^TA\mathbf{z}^{(0)}} \frac{77}{107},$$

and

$$\mathbf{x}^{(1)} = \mathbf{x}^{(0)} + \alpha_0\mathbf{z}^{(0)} = (197/428, -32/321)^T.$$

Solution 5.16 In the stationary case, $\rho(B_\alpha) = \min_\lambda|1 - \alpha\lambda|$, where λ are the eigenvalues of $P^{-1}A$. The optimal value of α is obtained solving the equation $|1 - \alpha\lambda_{min}| = |1 - \alpha\lambda_{max}|$, that is $1 - \alpha\lambda_{min} = -1 + \alpha\lambda_{max}$, which yields (5.39). Since,

$$\rho(B_\alpha) = 1 - \alpha\lambda_{min} \ \forall\alpha \leq \alpha_{opt},$$

for $\alpha = \alpha_{opt}$ we obtain (5.44).

Solution 5.17 In this case the matrix associated to the Leontieff model is not positive definite. Indeed, using the following instructions:

```
>> for i=1:20; for j=1:20; c(i,j)=i+j; end; end; A=eye(20)-c;
>>  min(eig(A))
ans =
 -448.5830
>> max(eig(A))
ans =
  30.5830
```

we can see that the minimum eigenvalue is a negative number and the maximum eigenvalue is a positive number. Therefore, the convergence of the gradient method is not guaranteed. However, since A is non singular, the given system is equivalent to the system $A^TA\mathbf{x} = A^T\mathbf{b}$, where A^TA is symmetric and positive definite. We solve the latter by the gradient method requiring that the norm of the residual be less than 10^{-10} and starting from the initial data $\mathbf{x}^{(0)} = \mathbf{0}^T$:

```
>> b = [1:20]';  aa=A'*A; b=A'*b; x0 = zeros(20,1);
>> [x,iter]=richardson(aa,b,x0,100,1.e-10);
```

The method converges in 15 iterations. A drawback of this approach is that the condition number of the matrix A^TA is, in general, larger than the condition number of A.

9.6 Chapter 6

Solution 6.1 A_1: the power method converges in 34 iterations to the value 2.00000000004989. A_2: starting from the same initial vector, the power method requires now 457 iterations to converge to the value 1.99999999990611. The slower convergence rate can be explained by observing that the two largest eigenvalues are very close one another. Finally, for the matrix A_3 the method doesn't converge since A_3 features two distinct eigenvalues (i and $-i$) of maximum modulus.

Solution 6.2 The Leslie matrix associated with the values in the table is

$$A = \begin{bmatrix} 0 & 0.5 & 0.8 & 0.3 \\ 0.2 & 0 & 0 & 0 \\ 0 & 0.4 & 0 & 0 \\ 0 & 0 & 0.8 & 0 \end{bmatrix}.$$

Using the power method we find $\lambda_1 \simeq 0.5353$. The normalized distribution of this population for different age intervals is given by the components of the corresponding unitary eigenvector, that is, $x_1 \simeq (0.8477, 0.3167, 0.2367, 0.3537)^T$.

Solution 6.3 We rewrite the initial guess as

$$\mathbf{y}^{(0)} = \beta^{(0)} \left(\alpha_1 \mathbf{x}_1 + \alpha_2 \mathbf{x}_2 + \sum_{i=3}^{n} \alpha_i \mathbf{x}_i \right),$$

with $\beta^{(0)} = 1/\|\mathbf{x}^{(0)}\|$. By calculations similar to those carried out in Section 6.1, at the generic step k we find:

$$\mathbf{y}^{(k)} = \gamma^k \beta^{(k)} \left(\alpha_1 \mathbf{x}_1 e^{ik\vartheta} + \alpha_2 \mathbf{x}_2 e^{-ik\vartheta} + \sum_{i=3}^{n} \alpha_i \frac{\lambda_i^k}{\gamma^k} \mathbf{x}_i \right).$$

The first two terms don't vanish and, due to the opposite sign of the exponents, the sequence of the $\mathbf{y}^{(k)}$ oscillates and cannot converge.

Solution 6.4 From the eigenvalue equation $A\mathbf{x} = \lambda \mathbf{x}$, we deduce $A^{-1}A\mathbf{x} = \lambda A^{-1}\mathbf{x}$, and therefore $A^{-1}\mathbf{x} = (1/\lambda)\mathbf{x}$.

Solution 6.5 The power method applied to the matrix A generates an oscillating sequence of approximations of the maximum modulus eigenvalue (see, Figure 9.8). This behavior is due to the fact that this eigenvalue is not unique.

Solution 6.6 To compute the eigenvalue of maximum modulus of A we use Program 9:

```
>> A=wilkinson(7);
>> x0=ones(7,1); tol=1.e-15; nmax=100;
>> [lambda,x,iter]=eigpower(A,tol,nmax,x0);
```

After 35 iterations we obtain `lambda=3.76155718183189`. To find the largest negative eigenvalue of A, we can use the power method with shift and, in particular, we can choose a shift equal to the largest positive eigenvalue that we have just computed. We find:

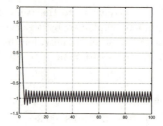

Fig. 9.8. The approximations of the maximum modulus eigenvalue of the matrix of Solution 6.5 computed by the power method

```
>> [lambda2,x,iter]=eigpower(A-lambda*eye(7),tol,nmax,x0);
>> lambda2+lambda
ans =
  -1.12488541976457
```

after `iter = 33` iterations. These results are satisfactory approximations of the largest (positive and negative) eigenvalues of A.

Solution 6.7 Since all the coefficients of A are real, eigenvalues occur in conjugate pairs. Note that in this situation conjugate eigenvalues must belong to the same Gershgorin circle. The matrix A presents 2 column circles isolated from the others (see Figure 9.9 on the left). Each of them must contain only one eigenvalue that must therefore be real. Then A admits at least 2 real eigenvalues.

Let us consider now the matrix B that admits only one column isolated circle (see Figure 9.9 on the right). Then, thanks to the previous consideration the corresponding eigenvalue must be real. The remaining eigenvalues can be either all real, or one real and 2 complex.

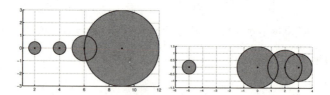

Fig. 9.9. On the left, column circles of the matrix A of Solution 6.7. On the right, column circles of the matrix B of Solution 6.7

Solution 6.8 The row circles of A feature an isolated circle of center 5 and radius 2 the maximum modulus eigenvalue must belong to. Therefore, we can set the value of the shift equal to 5. The comparison between the number of iterations and the computational cost of the power method with and without shift can be found using the following commands:

```
>> A=[5 0 1 -1; 0 2 0 -1/2; 0 1 -1 1; -1 -1 0 0];
>> flops(0); [lambda,x,iter]=eigpower(A,tol,nmax,x0); flops, iter
ans =
      2204
iter =
   34
>> flops(0); [lambda,x,iter]=invshift(A,tol,nmax,5,x0); flops, iter
ans =
      1082
iter =
   12
```

The power method with shift requires in this case a lower number of iterations (1 versus 3) and half the cost than the usual power method (also accounting for the extra time needed to compute the Gauss factorization of A off-line).

Solution 6.9 Using the qr command we have immediately:

```
>> A=[2 -1/2 0 -1/2; 0 4 0 2; -1/2 0 6 1/2; 0 0 1 9];
>> [Q,R]=qr(A)
Q =
  -0.9701    0.0073   -0.2389   -0.0411
        0   -0.9995   -0.0299   -0.0051
   0.2425    0.0294   -0.9557   -0.1643
        0         0   -0.1694    0.9855
R =
  -2.0616    0.4851    1.4552    0.6063
        0   -4.0018    0.1764   -1.9881
        0         0   -5.9035   -1.9426
        0         0         0    8.7981
```

To verify that RQ is similar to A, we observe that

$$Q^T A = Q^T QR = R$$

thanks to the orhogonality of Q. Thus $C = Q^T AQ = RQ$, since $Q^T = Q^{-1}$, and we conclude that C is similar to A.

Solution 6.10 A simple implementation of the proposed algorithm is:

```
function [A]=qrsimple(A,it)
for i = 1:it
 [Q,R]=qr(A);
 A = R*Q;
end
```

where it represents the number of iterations once the algorithm is stopped. For the given matrix, after 10 iterations we obtain the matrix $A^{(10)}$:

```
>> A10 =
    9.2090    1.1678    1.5237    0.0705
    0.1220    4.9228   -0.5739    0.0461
   -0.1732   -1.3095    4.8686   -0.6902
        0         0    0.0017    1.9996
```

while, after 50 iterations we obtain the matrix $A^{(50)}$:

```
>> A50 =
    9.1728   -0.8113    1.8631    0.1059
    0.0000    5.8003    0.6208    0.5914
   -0.0000   -0.0000    4.0268   -0.3526
        0         0    0.0000    2.0000
```

Note that the matrices $A^{(k)}$ are similar since $A^{(k+1)} = (Q^{(k)})^T A^{(k)} Q^{(k)}$ for $k \geq 0$ (see Solution 6.9). Therefore, we can guess that the sequence of matrices $A^{(k)}$ when $k \to \infty$ tends to an upper triangular matrix whose elements are the eigenvalues of A.

Solution 6.11 We can use the command `eig` in the following way: `[X,D]=eig(A)`, where X is the matrix whose columns are the unit eigenvectors of A and D is a diagonal matrix whose elements are the eigenvalues of A. For the matrices A and B of Exercise 6.7 we should execute the following instructions:

```
>> A=[2 -1/2 0 -1/2; 0 4 0 2; -1/2 0 6 1/2; 0 0 1 9];
>> eig(A)
ans =
    2.0000
    5.8003
    9.1728
    4.0268
>> B=[-5 0 1/2 1/2; 1/2 2 1/2 0; 0 1 0 1/2; 0 1/4 1/2 3];
>> eig(B)
ans =
   -4.9921
    2.1666
   -0.3038
    3.1292
```

9.7   Chapter 7

Solution 7.1 Let us approximate the exact solution $y(t) = \frac{1}{2}[e^t - \sin(t) - \cos(t)]$ of the Cauchy problem (7.43) by the forward Euler method using different values of h: $1/2, 1/4, 1/8, \ldots, 1/512$. The associated error is computed by the following instructions:

```
>> y0=0; f=inline('sin(t)+y','t','y'); y='0.5*(exp(t)-sin(t)-cos(t))';
```

```
>> tspan=[0 1]; N=2; for k=1:10
[tt,u]=feuler(f,tspan,y0,N);t=tt(end);e(k)=abs(u(end)-eval(y));N=2*N;end
>> e
e =
```
  Columns 1 through 7
    0.4285   0.2514   0.1379   0.0725   0.0372   0.0189   0.0095
  Columns 8 through 10
    0.0048   0.0024   0.0012

Now we apply formula (1.11) to estimate the order of convergence:

```
>> p=log(abs(e(1:end-1)./e(2:end)))/log(2)
p =
```
  Columns 1 through 7
    0.7696   0.8662   0.9273   0.9620   0.9806   0.9902   0.9951
  Columns 8 through 9
    0.9975   0.9988

As expected the order of convergence is one. With the same instructions (substituting the program **feuler** with the program **beuler**) we obtain an estimate of the convergence order of the backward Euler method:

```
>> p=log(abs(e(1:end-1)./e(2:end)))/log(2)
p =
```
  Columns 1 through 7
    1.5199   1.1970   1.0881   1.0418   1.0204   1.0101   1.0050
  Columns 8 through 9
    1.0025   1.0012

Solution 7.2 The numerical solution of the given Cauchy problem by the forward Euler method can be obtained as follows:

```
>> tspan=[0 1]; N=100;f=inline('-t*exp(-y)','t','y');y0=0;
>> [t,u]=feuler(f,tspan,y0,N);
```

To compute the number of exact significant digits we can estimate the constants L and M which appear in (7.11). Note that, since $f(t, y(t)) < 0$ in the given interval, $y(t) = \log(1 - t^2/2)$ is a monotonically decreasing function, vanishing at $t = 0$. Since f is continuous together with its first derivative, we can approximate L as $L = \max_{0 \le t \le 1} |L(t)|$ with $L(t) = \partial f/\partial y = te^{-y}$. Note that $L(0) = 0$ and $L(t) > 0$ for all $t \in (0, 1]$. Thus, $L = e$.

Similarly, in order to compute $M = \max_{0 \le t \le 1} |y''(t)|$ with $y'' = -e^{-y} - t^2 e^{-2y}$, we can observe that this function has its maximum at $t = 1$, and then $M = e + e^2$. From (7.11) we deduce

$$|u_{100} - y(1)| \le \frac{e^L - 1}{L} \frac{M}{200} = 0.26.$$

Therefore, there is no guarantee that more than one significant digit be exact. Indeed, we find u(end)=-0.6785, while the exact solution at $t = 1$ is $y(1) = -0.6931$.

Solution 7.3 The iteration function is $\phi(u) = u_n - ht_{n+1}e^{-u}$ and the fixed-point iteration converges if $|\phi'(u)| < 1$. This property is ensured if $h(t_0 + (n+1)h) < e^u$. If we substitute u with the exact solution, we can provide an *a priori* estimate of the value of h. The most restrictive situation occurs when $u = -1$ (see Solution 7.2). In this case the solution of the inequality $(n+1)h^2 < e^{-1}$ is $h < \sqrt{e^{-1}/(n+1)}$.

Solution 7.4 We repeat the same set of instructions of Solution 7.1, however now we use the program cranknic (Program 14) instead of feuler. According to the theory, we obtain the following result that shows second-order convergence:

```
>> p=log(abs(e(1:end-1)./e(2:end)))/log(2)
p =
  Columns 1 through 7
    2.0379   2.0092   2.0023   2.0006   2.0001   2.0000   2.0000
  Columns 8 through 9
    2.0000   2.0000
```

Solution 7.5 Consider the integral formulation of the Cauchy problem (7.4) in the interval $[t_n, t_{n+1}]$:

$$
y(t_{n+1}) - y(t_n) = \int_{t_n}^{t_{n+1}} f(\tau, y(\tau))d\tau
$$
$$
\simeq \frac{h}{2}[f(t_n, y(t_n)) + f(t_{n+1}, y(t_{n+1}))],
$$

where we have approximated the integral by the trapezoidal formula (4.17). By setting $u_0 = y(t_0)$ and replacing $y(t_n)$ by the approximate value u_n and the symbol $\simeq$ by $=$, we obtain

$$
u_{n+1} = u_n + \frac{h}{2}[f(t_n, u_n) + f(t_{n+1}, u_{n+1})], \qquad \forall n \geq 0,
$$

which is the Crank-Nicolson method.

Solution 7.6 Set $h\lambda = x + iy$, so that $|1 + h\lambda|^2 = (1 + x)^2 + y^2$. Then, the assumption $|1 + h\lambda| < 1$ is equivalent to $(1 + x)^2 + y^2 < 1$. It follows that the region of absolute stability for the forward Euler method is the unit circle in the complex plane (see Figure 9.11).

Solution 7.7 Thanks to the result found in Solution 7.6, we must impose the limitation $|1 - h + ih| < 1$, which yields $0 < h < 1$.

Solution 7.8 Let us rewrite the Heun method in the following (Runge-Kutta like) form:

$$
u_{n+1} = u_n + \frac{1}{2}(k_1 + k_2), \quad k_1 = hf(t_n, u_n), \quad k_2 = hf(t_{n+1}, u_n + k_1). \quad (9.4)
$$

We have $h\tau_{n+1}(h) = y(t_{n+1}) - y(t_n) - (\widehat{k}_1 + \widehat{k}_2)/2$, with $\widehat{k}_1 = hf(t_n, y(t_n))$ and $\widehat{k}_2 = hf(t_{n+1}, y(t_n) + \widehat{k}_1)$. Therefore, the method is consistent since

$$\lim_{h \to 0} \tau_{n+1} = y'(t_n) - \frac{1}{2}[f(t_n, y(t_n)) + f(t_n, y(t_n))] = 0.$$

The Heun method is implemented in Program 19. Using this program, we can verify the order of convergence as in Solution 7.1. By the following instructions, we find that the Heun method is second-order with respect to h

```
>>  p=log(abs(e(1:end-1)./e(2:end)))/log(2)
p =
  Columns 1 through 7
    1.7642    1.8796    1.9398    1.9700    1.9851    1.9925    1.9963
  Columns 8 through 9
    1.9981    1.9991
```

Program 19 - rk2: the Heun method

```
function [t,u]=rk2(odefun,tspan,y0,Nh,varargin)
h=(tspan(2)-tspan(1)-t0)/Nh; tt=[tspan(1):h:tspan(2)]; u(1)=y0;
for s=tt(1:end-1)
    t = s;  y = u(end); k1=h*feval(odefun,t,y,varargin{:});
    t = t + h; y = y + k1; k2=h*feval(odefun,t,y,varargin{:});
    u = [u, u(end) + 0.5*(k1+k2)];
end
t=tt;
```

Solution 7.9 Applying the method (9.4) to the model problem (7.19) we obtain $k_1 = h\lambda u_n$ and $k_2 = h\lambda u_n(1 + h\lambda)$. Therefore $u_{n+1} = u_n[1 + h\lambda + (h\lambda)^2/2] = u_n p_2(h\lambda)$. To ensure absolute stability we must require that $|p_2(h\lambda)| < 1$, which is equivalent to $0 < p_2(h\lambda) < 1$, since $p_2(h\lambda)$ is positive. Solving the latter inequality, we obtain $-2 < h\lambda < 0$, that is, $h < 2/|\lambda|$.

Solution 7.10 Note that

$$u_n = u_{n-1}(1 + h\lambda_{n-1}) + hr_{n-1}.$$

Then proceed recursively on n.

Solution 7.11 The inequality (7.29) follows from (7.28) by setting

$$\varphi(\lambda) = \left|1 + \frac{1}{\lambda}\right| + \left\|\frac{1}{\lambda}\right\|.$$

The conclusion follows easily.

Solution 7.12 From (7.26) we have

$$|z_n - u_n| \leq \rho_{max} a^n + h\rho_{max} \sum_{k=0}^{n-1} \delta(h)^{n-k-1}.$$

The result follows using (7.27).

Solution 7.13 We have

$$h\tau_{n+1}(h) = y(t_{n+1}) - y(t_n) - \frac{1}{6}(\widehat{k}_1 + 4\widehat{k}_2 + \widehat{k}_3),$$

$$\widehat{k}_1 = hf(t_n, y(t_n)), \quad \widehat{k}_2 = hf(t_n + \tfrac{h}{2}, y(t_n) + \tfrac{\widehat{k}_1}{2}),$$

$$\widehat{k}_3 = hf(t_{n+1}, y(t_n) + 2\widehat{k}_2 - \widehat{k}_1).$$

This method is consistent since

$$\lim_{h\to 0} \tau_{n+1} = y'(t_n) - \frac{1}{6}[f(t_n, y(t_n)) + 4f(t_n, y(t_n)) + f(t_n, y(t_n))] = 0.$$

This method is an explicit Runge-Kutta method of order 3 and is implemented in Program 20. As in Solution 7.8, we can derive an estimate of its order of convergence by the following instructions:

```
>> p=log(abs(e(1:end-1)./e(2:end)))/log(2)
p =
  Columns 1 through 7
    2.7306    2.8657    2.9330    2.9666    2.9833    2.9916    2.9958
  Columns 8 through 9
    2.9979    2.9990
```

Solution 7.14 From Solution 7.9 we obtain the relation

$$u_{n+1} = u_n[1 + h\lambda + \frac{1}{2}(h\lambda)^2 + \frac{1}{6}(h\lambda)^3] = u_n p_3(h\lambda).$$

By inspection of the graph of p_3, obtained with the instruction

```
>> c=[1/6 1/2 1 1]; z=[-3:0.01:1]; p=polyval(c,z); plot(z,abs(p))
```

we deduce that $|p_3(h\lambda)| < 1$ for $-2.5 < h\lambda < 0$.

Program 20 - rk3: Explicit Runge-Kutta method of order 3

```
function [t,u]=rk3(odefun,tspan,y0,Nh,varargin)
h=(tspan(2)-tspan(1))/Nh; tt=[tspan(1):h:tspan(2)]; u(1)=y0;
for s=tt(1:end-1)
t = s;        y = u(end);        k1=h*feval(odefun,t,y,varargin{:});
t = t + h*0.5; y = y + 0.5*k1;   k2=h*feval(odefun,t,y,varargin{:});
```

```
t = s + h;     y = u(end) + 2*k2-k1; k3=h*feval(odefun,t,y,varargin{:});
u = [u, u(end) + (k1+4*k2+k3)/6];
end
t=tt;
```

Solution 7.15 The method (7.46) applied to the model problem (7.19) gives the equation $u_{n+1} = u_n(1 + h\lambda + (h\lambda)^2)$. From the graph of $1 + z + z^2$ with $z = h\lambda$, we deduce that the method is absolutely stable if $-1 < h\lambda < 0$.

Solution 7.16 To solve Problem 7.1 with the given values, we repeat the following instructions with N=10 and N=20:

```
>> f=inline('-1.68*10^(-9)*y^4+2.6880','t','y');
>> [t,uc]=cranknic(f,[0,200],180,N);
>> [t,u]=predcor(f,[0 200],180,N,'feonestep','cnonestep');
```

The graphs of the computed solutions are shown in Figure 9.10. The solutions obtained by the Crank-Nicolson method are more accurate than those obtained by the Heun method.

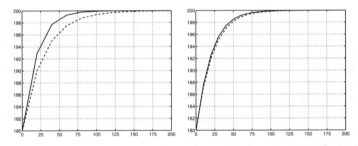

Fig. 9.10. Computed solutions with $h = 20$ (left) and $h = 10$ (right) for the Cauchy problem of Solution 7.16: in the continuous line, the solutions computed by the Crank-Nicolson method, in the dashed lines those computed by the Heun method

Solution 7.17 Heun method applied to the model problem (7.19), gives

$$u_{n+1} = u_n \left(1 + h\lambda + \frac{1}{2}h^2\lambda^2\right).$$

In the complex plane the boundary of its region of absolute stability satisfies $|1+h\lambda+h^2\lambda^2/2|^2 = 1$, having set $h\lambda = x+iy$. This equation is satisfied by the numbers (x, y) such that $f(x, y) = x^4 + y^4 + 2x^2y^2 + 4x^3 + 4xy^2 + 8x^2 + 8x = 0$. We can represent this curve as the level curve $f(x, y) = z$ (corresponding to the level $z = 0$). This can be done by means of the following instructions:

```
>> f='x.^4+y.^4+2*(x.^2).*(y.^2)+4*x.*y.^2+4*x.^3+8*x.^2+8*x';
>> [x,y]=meshgrid([-2.1:0.1:0.1],[-2:0.1:2]);
>> contour(x,y,eval(f),[0 0])
```

meshgrid The command meshgrid draws in the rectangle $[-2.1, 0.1] \times [-2, 2]$ a grid with 23 equispaced nodes in the x-direction, and 41 equispaced nodes in the

contour y-direction. With the command contour we plot the level curve of $f(x,y)$ (evaluated with the command eval(f)) corresponding to the value $z = 0$ (made precise in the input vector [0 0] of contour). In Figure 9.11 the continuous line delimitates the region of absolute stability of the Heun method. This region is larger than the corresponding region of the forward Euler method (which corresponds to the interior of the dashed circle). Both curves are tangent to the imaginary axis at the origin $(0, 0)$.

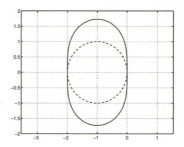

Fig. 9.11. Boundaries of the regions of absolute stability for the Heun method (continuous line) and the forward Euler method (dashed line). The corresponding regions lie at the interior of the boundaries

Solution 7.18 We use the following instructions:

```
>> tspan=[0 1]; y0=0; f=inline('cos(2*y)','t','y');
>> y='0.5*asin((exp(4*t)-1)./(exp(4*t)+1))';
>> N=2; for k=1:10
   [tt,u]=predcor(f,tspan,y0,N,'feonestep','cnonestep');
   t=tt(end); e(k)=abs(u(end)-eval(y)); N=2*N; end
>>  p=log(abs(e(1:end-1)./e(2:end)))/log(2)
p =
  Columns 1 through 7
    2.4733   2.2507   2.1223   2.0601   2.0298   2.0148   2.0074
  Columns 8 through 9
    2.0037   2.0018
```

As expected, we find that the order of convergence of the method is 2. However, the computational cost is comparable with that of the forward Euler method, which is first-order accurate only.

Solution 7.19 The second-order differential equation of this exercise is equivalent to the following first-order system:

$$x' = z, \quad z' = -5z - 6x,$$

with $x(0) = 1$, $z(0) = 0$. We use the Heun method as follows:

```
>> tspan=[0 5]; y0=[1 0];
>> [tt,u]=predcor('fspring',tspan,y0,N,'feonestep','cnonestep');
```

where N is the number of nodes and `fspring.m` is the following function:

```
function y=fspring(t,y)
b=5; k=6;
yy=y; y(1)=yy(2); y(2)=-b*yy(2)-k*yy(1);
```

In Figure 9.12 we show the graphs of the two components of the solution, computed with N=20,40 and compare them with the graph of the exact solution $x(t) = 3e^{-2t} - 2e^{-3t}$ and that of its first derivative.

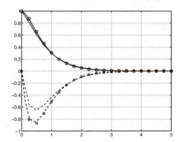

Fig. 9.12. Approximations of $x(t)$ (continuous line) and $x'(t)$ (dashed line) computed with N=20 (thin line) and N=40 (thick line). Small circles and squares refer to the exact functions $x(t)$ and $x'(t)$, respectively

Solution 7.20 The second-order system of differential equations is reduced to the following first-order system:

$$\begin{cases} x' = z, \\ y' = v, \\ z' = 2\omega \sin(\Psi) - k^2 x, \\ v' = -2\omega \sin(\Psi)z - k^2 y. \end{cases} \tag{9.5}$$

If we suppose that the pendulum at the initial time $t_0 = 0$ is at rest in the position $(1, 0)$, the system (9.5) must be given the following initial conditions:

$$x(0) = 1, \quad y(0) = 0, \quad z(0) = 0, \quad v(0) = 0.$$

Setting $\Psi = \pi/4$, which is the average latitude of the Northern Italy, we use the forward Euler method as follows:

>> [t,y]=feuler('ffocault',[0 300],[1 0 0 0],Nh);

where Nh is the number of steps and ffocault.m is the following function:

```
function y=ffocault(t,y)
l=20;   k2=9.8/l;   psi=pi/4; omega=7.29*1.e-05;
yy=y;   y(1)=yy(3);  y(2)=yy(4);
y(3)=2*omega*sin(psi)*yy(4)-k2*yy(1);
y(4)=-2*omega*sin(psi)*yy(3)-k2*yy(2);
```

By some numerical experiments we conclude that the forward Euler method cannot produce acceptable solutions for this problem even for very small h. For instance, on the left of Figure 9.13 we show the graph, in the phase plane (x, y), of the motion of the pendulum computed with N=30000, that is, $h = 1/100$. As expected, the rotation plane changes with time, but also the amplitude of the oscillations increases. Similar results can be obtained for smaller h and using the Heun method. In fact, the model problem corresponding to the problem at hand has a coefficient λ that is purely imaginary. The corresponding solution (a sinusoid) is bounded for t that tends to infinity, however it doesn't tend to zero.

Unfortunately, both the forward Euler and Heun methods feature a region of absolute stability that doesn't include any point of the imaginary axis (with the exception of the origin). Thus, to ensure the absolute stability one should choose the prohibited value $h = 0$.

To get an acceptable solution we should use a method whose region of absolute stability includes a portion of the imaginary axis. This is the case, for instance, for the adaptive Runge-Kutta method of order 3, implemented ode23 in the MATLAB function ode23. We can invoke it by the following command:

>> [t,u]=ode23('ffocault',[0 300],[1 0 0 0]);

In Figure 9.13 (right) we show the solution obtained using only 1022 integration steps. Note that the numerical solution is in good agreement with the analytical one.

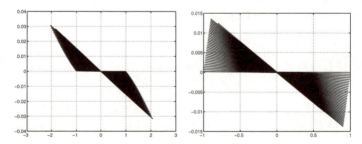

Fig. 9.13. Trajectories on the phase plane for the Focault pendulum of Solution 7.20 computed by the forward Euler method (left) and the third-order adaptive Runge-Kutta method (right)

9.8   Chapter 8

Solution 8.1 We can verify directly that $\mathbf{x}^T A \mathbf{x} > 0$ for all $\mathbf{x} \neq \mathbf{0}$. Indeed,

$$
[x_1\ x_2\ \ldots\ x_{N-1}\ x_N]
\begin{bmatrix}
2 & -1 & 0 & \cdots & 0 \\
-1 & 2 & \ddots & & \vdots \\
0 & \ddots & \ddots & -1 & 0 \\
\vdots & & -1 & 2 & -1 \\
0 & \cdots & 0 & -1 & 2
\end{bmatrix}
\begin{bmatrix}
x_1 \\ x_2 \\ \vdots \\ x_{N-1} \\ x_N
\end{bmatrix}
$$

$$
= 2x_1^2 - 2x_1x_2 + 2x_2^2 - 2x_2x_3 + \ldots - 2x_{N-1}x_N + 2x_N^2.
$$

The last expression is equivalent to $(x_1 - x_2)^2 + \ldots + (x_{N-1} - x_N)^2 + x_1^2 + x_N^2$, which is, positive provided that at least one x_i is non-null.

Solution 8.2 We verify that $A\mathbf{q}_j = \lambda_j \mathbf{q}_j$. Computing the matrix-vector product $\mathbf{w} = A\mathbf{q}_j$ and requiring that $\mathbf{w}$ is equal to the vector $\lambda_j \mathbf{q}_j$, we find:

$$
\begin{cases}
2\sin(j\theta) - \sin(2j\theta) = 2(1 - \cos(j\theta))\sin(j\theta), \\
-\sin(jk\theta) + 2\sin(j(k+1)\theta) - \sin(j(k+2)\theta) = 2(1 - \cos(j\theta))\sin(2j\theta), \\
\qquad k = 1, \ldots, N-2 \\
2\sin(Nj\theta) - \sin((N-1)j\theta) = 2(1 - \cos(j\theta))\sin(Nj\theta).
\end{cases}
$$

The first equation is an identity since $\sin(2j\theta) = 2\sin(j\theta)\cos(j\theta)$. The other equations can be simplified since

$$
\sin(jk\theta) = \sin((k+1)j\theta)\cos(j\theta) - \cos((k+1)j\theta)\sin(j\theta),
$$

$$
\sin(j(k+2)\theta) = \sin((k+1)j\theta)\cos(j\theta) + \cos((k+1)j\theta)\sin(j\theta).
$$

Since A is symmetric and positive definite, its condition number is $K(A) = \lambda_{max}/\lambda_{min}$, that is, $K(A) = \lambda_1/\lambda_N = (1 - \cos(N\pi/(N+1)))/(1 - \cos(\pi/(N+1)))$. Using the Taylor expansion of order 2 of the cosine function, we obtain $K(A) \simeq N^2$, that is, $K(A) \simeq h^{-2}$.

Solution 8.3 We note that

$$
u(\bar{x} + h) = u(\bar{x}) + hu'(\bar{x}) + \frac{h^2}{2}u''(\bar{x}) + \frac{h^3}{6}u'''(\bar{x}) + \frac{h^4}{24}u^{(4)}(\xi_+),
$$

$$
u(\bar{x} - h) = u(\bar{x}) - hu'(\bar{x}) + \frac{h^2}{2}u''(\bar{x}) - \frac{h^3}{6}u'''(\bar{x}) + \frac{h^4}{24}u^{(4)}(\xi_-),
$$

where $\xi_+ \in (x, x+h)$ e $\xi_- \in (x-h, x)$. Summing the two expression we obtain

$$
u(\bar{x} + h) + u(\bar{x} - h) = 2u(\bar{x}) + h^2 u''(\bar{x}) + \frac{h^4}{24}(u^{(4)}(\xi_+) + u^{(4)}(\xi_-)),
$$

which is the desired property.

Solution 8.4 The matrix is again tridiagonal with entries $a_{i,i-1} = -1 - h\frac{\delta}{2}$, $a_{ii} = 2 + h^2\gamma$, $a_{i,i+1} = -1 + h\frac{\delta}{2}$. The right-hand side, accounting for the boundary conditions, becomes $\mathbf{f} = (f(x_1) + \alpha(1+h\delta/2)/h^2, f(x_2), \ldots, f(x_{N-1}), f(x_N) + \beta(1 - h\delta/2)/h^2)^T$.

Solution 8.5 With the following instructions we compute the corresponding solutions to the three given values of h:

```
>> fbvp=inline('1+sin(4*pi*x)','x');
>> [z,uh10]=bvp(0,1,9,0,0.1,fbvp,0,0);
>> [z,uh20]=bvp(0,1,19,0,0.1,fbvp,0,0);
>> [z,uh40]=bvp(0,1,39,0,0.1,fbvp,0,0);
```

Since we don't know the exact solution, to estimate the convergence order we compute an approximate solution on a very fine grid (for instance $h = 1/1000$), then we use this latter as a surrogate for the exact solution. We find:

```
>> [z,uhex]=bvp(0,1,999,0,0.1,fbvp,0,0);
>> max(abs(uh9-uhex(1:100:end)))
ans =
   8.6782e-04
>> max(abs(uh19-uhex(1:50:end)))
ans =
   2.0422e-04
>> max(abs(uh39-uhex(1:25:end)))
ans =
   5.2789e-05
```

Halving h, the error is divided by 4, proving that the convergence order with respect to h is 2.

Solution 8.6 To find the largest h_{crit} which ensures a monotonic solution (as the analytical one) we execute the following cycle:

```
>> fbvp=inline('1+0.*x','x'); for k=3:1000
   [z,uh]=bvp(0,1,k,100,0,fbvp,0,1); if sum(diff(uh)>0)==length(uh)-1,
   break, end, end
```

We let $h(= 1/(k+1))$ vary till the forward incremental ratios of the numerical solution uh are all positive. Then we compute the vector diff(uh) whose components are 1 if the corresponding incremental ratio is positive, 0 otherwise. If the sum of all components equals the vector length of uh diminished by 1, then all incremental ratios are positive. The cycle stops when k=49, that is, when $h = 1/500$ if $\delta = 1000$, and when $h = 1/1000$ if $\delta = 2000$. We can therefore guess that one should require $h < 2/\delta = h_{crit}$ in order to get a monotonically increasing numerical solution. Indeed, this restriction on h is precisely what can be proven theoretically (see, for instance, [QV94]). In Figure 9.14 we show the numerical solutions obtained when $\delta = 100$ for two values of h.

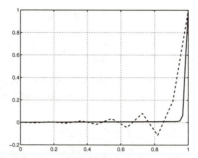

Fig. 9.14. Numerical solution for Problem 8.6 obtained for $h = 1/10$ (dashed line) and $h = 1/60$ (continuous line)

Solution 8.7 We should modify the Program 17 in order to impose Neumann boundary conditions. In the Program 21 we show one possible implementation.

Program 21 - neumann: approximation of a Neumann boundary problem

```
function [x,uh]=neumann(a,b,N,delta,gamma,bvpfun,ua,ub,varargin)
h = (b-a)/(N+1); x = [a:h:b]; e = ones(N+2,1);
A = spdiags([-e-0.5*h*delta 2*e+gamma*h^2 -e+0.5*h*delta], -1:1, N+2, N+2);
f = h^2*feval(bvpfun,'x',varargin{:}); f=f';
A(1,1)=-3/2*h; A(1,2)=2*h; A(1,3)=-1/2*h; f(1)=h^2*ua;
A(N+2,N+2)=3/2*h; A(N+2,N+1)=-2*h; A(N+2,N)=1/2*h; f(N+2)=h^2*ub;
uh = A\f;
```

Solution 8.8 The trapezoidal integration formula, used on the two subintervals I_{k-1} and I_k, produces the following approximation

$$\int_{I_{k-1} \cup I_k} f(x)\varphi_k(x)\, dx \simeq \frac{h}{2}f(x_k) + \frac{h}{2}f(x_k) = hf(x_k),$$

since $\varphi_k(x_j) = \delta_{jk}$, $\forall j, k$. Thus, we obtain the same right-hand side of the finite difference method.

Solution 8.9 We have $\nabla\phi = (\partial\phi/\partial x, \partial\phi/\partial y)^T$ and therefore $\text{div}\nabla\phi = \partial^2\phi/\partial x^2 + \partial^2\phi/\partial y^2$, that is, the Laplacian of ϕ.

Solution 8.10 To compute the temperature at the center of the plate, we solve the corresponding Poisson problem for various values of $\Delta_x = \Delta_y$, using the following instructions:

```
>> k=0; fun=inline('25','x','y'); bound=inline('(x==1)','x','y');
>> for N = [10,20,40,80,160],
   [u,x,y]=poissonfd(0,0,1,1,N,N,fun,bound);
   k=k+1; uc(k) = u(N/2+1,N/2+1); end
```

The components of the vector uc are the values of the computed temperature
at the center of the plate as the step-size h of the grid decreases. We have

```
>> uc
     2.0168    2.0616    2.0789    2.0859    2.0890
```

We can therefore conclude that at the center of the plate the temperature is
approximatively 2.08 C. In Figure 9.15 we show the level-curve of the temperature for two different values of h.

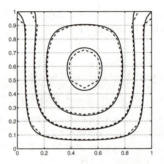

Fig. 9.15. The level-curve of the computed temperature for $\Delta_x = \Delta_y = 1/10$
(dashed lines) and for $\Delta_x = \Delta_y = 1/80$ (continuous lines)

Bibliography

[Arn73]   Arnold V. (1973) *Ordinary Differential Equations*. The MIT Press, Cambridge, Massachusetts.

[Atk89]   Atkinson K. (1989) *An Introduction to Numerical Analysis*. John Wiley, New York.

[Axe94]   Axelsson O. (1994) *Iterative Solution Methods*. Cambridge University Press, New York.

[BB96]   Brassard G. and Bratley P. (1996) *Fundamentals of Algorithmics, 1/e*. Prentice Hall, New York.

[BM92]   Bernardi C. and Maday Y. (1992) *Approximations Spectrales des Problémes aux Limites Elliptiques*. Springer-Verlag, Paris.

[Bra97]   Braess D. (1997) *Finite Elements: Theory, Fast Solvers and Applications in Solid Mechanics*. Cambridge University Press, Cambridge.

[BS01]   Babuska I. and Strouboulis T. (2001) *The Finite Element Method and its Reliability*. Oxford University Press.

[But87]   Butcher J. (1987) *The Numerical Analysis of Ordinary Differential Equations: Runge-Kutta and General Linear Methods*. Wiley, Chichester.

[CHQZ88]   Canuto C., Hussaini M. Y., Quarteroni A., and Zang T. A. (1988) *Spectral Methods in Fluid Dynamics*. Springer, New York.

[Dav63]   Davis P. (1963) *Interpolation and Approximation*. Blaisdell Pub., New York.

[DB02]   Deuflhard P. and Bornemann F. (2002) *Scientific Computing with Ordinary Differential Equations*. Springer-Verlag, New York.

[DD95]   Davis T. and Duff I. (1995) *A Combined Unifrontal/Multifrontal Method for Unsymmetric Sparse Matrices*. TR-95-020. University of Florida.

[Die93]   Dierckx P. (1993) *Curve and Surface Fitting with Splines*. Claredon Press, New York.

[DL92]   DeVore R. and Lucier J. (1992) Wavelets. *Acta Numerica* pages 1–56.

[DR75]   Davis P. and Rabinowitz P. (1975) *Methods of Numerical Integration*. Academic Press, New York.

[DS83]   Dennis J. and Schnabel R. (1983) *Numerical Methods for Unconstrained Optimization and Nonlinear Equations*. Prentice-Hall, Englewood Cliffs, New York.

[dV89]   der Vorst H. V. (1989) High Performance Preconditioning. *SIAM J. Sci. Stat. Comput.* 10: 1174–1185.

[EEHJ96]   Eriksson K., Estep D., Hansbo P., and Johnson C. (1996) *Computational Differential Equations*. Cambridge Univ. Press, Cambridge.

[EKH02]   Etter D., Kuncicky D., and Hull D. (2002) *Introduction to MATLAB 6*. Prentice Hall.

[Fun92]   Funaro D. (1992) *Polynomial Approximation of Differential Equations*. Springer-Verlag, Berlin.

[Gau97]   Gautschi W. (1997) *Numerical Analysis. An Introduction*. Birkhäuser, Berlin.

[GL89]   Golub G. and Loan C. V. (1989) *Matrix Computations*. The John Hopkins Univ. Press, Baltimore and London.

[Hac85]   Hackbush W. (1985) *Multigrid Methods and Applications*. Springer-Verlag, Berlin.

[Hac94]   Hackbush W. (1994) *Iterative Solution of Large Sparse Systems of Equations*. Springer-Verlag, New York.

[HH00]   Higham D. and Higham N. (2000) *MATLAB Guide*. SIAM, Philadelphia.

[Hig96]   Higham N. (1996) *Accuracy and Stability of Numerical Algorithms*. SIAM Publications, Philadelphia, PA.

[Hir88]   Hirsh C. (1988) *Numerical Computation of Internal and External Flows*, volume 1. John Wiley and Sons, Chichester.

[HLR01]   Hunt B., Lipsman R., and Rosenberg J. (2001) *A guide to MATLAB: for Beginners and Experienced Users* . Cambridge University Press.

[IK66]   Isaacson E. and Keller H. (1966) *Analysis of Numerical Methods*. Wiley, New York.

[Iro70]   Irons B. (1970) A Frontal Solution Program for Finite Element Analysis. *Int. J. for Numer. Meth. in Engng.* 2: 5–32.

[Krö98]   Kröner D. (1998) *Finite volume schemes in multidimensions*. Pitman Res. Notes Math. Ser., 380, Longman, Harlow.

[KS99]   Karniadakis G. and Sherwin S. (1999) *Spectral/hp Element Methods for CFD*. Oxford University Press.

[Lam91]   Lambert J. (1991) *Numerical Methods for Ordinary Differential Systems*. John Wiley and Sons, Chichester.

[Lan03]   Langtangen H. (2003) *Computational Partial Differential Equations: Numerical Methods and Diffpack Programming*. Springer-Verlag, Heidelberg.

[LeV02]   LeVeque R. (2002) *Finite Volume Methods for Hyperbolic Problems*. Cambridge University Press, Cambridge.

[Mei67]   Meinardus G. (1967) *Approximation of Functions: Theory and Numerical Methods*. Springer-Verlag, New York.

248   Bibliography

[Pan92]   Pan V. (1992) Complexity of Computations with Matrices
             and Polynomials. *SIAM Review* 34: 225–262.

[PBP02]   Prautzsch H., Boehm W., and Paluszny M. (2002) *Bezier
             and B-Spline Techniques.* Springer-Verlag.

[PdDKÜK83]   Piessens R., de Doncker-Kapenga E., Üeberhuber C., and
             Kahaner D. (1983) *QUADPACK: A Subroutine Package
             for Automatic Integration.* Springer-Verlag, Berlin and
             Heidelberg.

[QSS00]   Quarteroni A., Sacco R., and Saleri F. (2000) *Numerical
             Mathematics*, volume 37 of *Texts in Applied Mathematics.*
             Springer-Verlag, New York.

[QV94]   Quarteroni A. and Valli A. (1994) *Numerical Approxima-
             tion of Partial Differential Equations.* Springer, Berlin
             and Heidelberg.

[RR85]   Ralston A. and Rabinowitz P. (1985) *A First Course in
             Numerical Analysis.* McGraw-Hill, Singapore. 7h print-
             ing.

[Saa92]   Saad Y. (1992) *Numerical Methods for Large Eigenvalue
             Problems.* Halstead Press, New York.

[Saa96]   Saad Y. (1996) *Iterative Methods for Sparse Linear Sys-
             tems.* PWS Publishing Company, Boston.

[SR97]   Shampine L. F. and Reichelt M. W. (1997) The MATLAB
             ODE Suite. *SIAM J. Sci. Comput.* 18: 1–22.

[TW98]   Tveito A. and Winther R. (1998) *Introduction to Par-
             tial Differential Equations. A Computational Approach.*
             Springer Verlag.

[Üeb97]   Üeberhuber C. (1997) *Numerical Computation: Methods,
             Software, and Analysis.* Springer-Verlag, Berlin Heidel-
             berg.

[Urb02]   Urban K. (2002) *Wavelets in Numerical Simulation.*
             Springer Verlag.

[Wes91]   Wesseling P. (1991) *An Introduction to Multigrid Methods.*
             John Wiley and Sons, Chichester.

[Wil65]   Wilkinson J. (1965) *The Algebraic Eigenvalue Problem.*
             Clarendon Press, Oxford.

Index of MATLAB Programs

Index

Editorial Policy

§1. Textbooks on topics in the field of computational science and engineering will be considered. They should be written for courses in CSE education. Both graduate and undergraduate textbooks will be published in TCSE. Multidisciplinary topics and multidisciplinary teams of authors are especially welcome.

§2. Format: Only works in English will be considered. They should be submitted in camera-ready form according to Springer-Verlag's specifications. Electronic material can be included if appropriate. Please contact the publisher. Technical instructions and/or TₑX macros are available via
http://www.springer.de/math/authors/help-momu.html.

§3. Those considering a book which might be suitable for the series are strongly advised to contact the publisher or the series editors at an early stage. Addresses are given on the next page.

General Remarks

TCSE books are printed by photo-offset from the master-copy delivered in camera-ready form by the authors. For this purpose Springer-Verlag provides technical instructions for the preparation of manuscripts. See also *Editorial Policy*.

Careful preparation of manuscripts will help keep production time short and ensure a satisfactory appearance of the finished book.

The following terms and conditions hold:

Regarding free copies and royalties, the standard terms for Springer mathematics monographs and textbooks hold. Please write to Peters@springer.de for details.

Authors are entitled to purchase further copies of their book and other Springer books for their personal use, at a discount of 33,3 % directly from Springer-Verlag.

Series Editors

Texts in Computational Science and Engineering

Vol. 1   H. P. Langtangen, *Computational Partial Differential Equations.* Numerical Methods and Diffpack Programming. 2nd Edition 2003. XXVI, 855 pp. Hardcover. ISBN 3-540-43416-X

Vol. 2   A. Quarteroni, F. Saleri, *Scientific Computing with MATLAB.* 2003.   IX, 257 pp. Hardcover. ISBN 3-540-44363-0

Lecture Notes in Computational Science and Engineering

Vol. 1   D. Funaro, *Spectral Elements for Transport-Dominated Equations.*   1997. X, 211 pp. Softcover. ISBN 3-540-62649-2

Vol. 2   H. P. Langtangen, *Computational Partial Differential Equations.* Numerical Methods and Diffpack Programming.   1999. XXIII, 682 pp. Hardcover. ISBN 3-540-65274-4

Vol. 3   W. Hackbusch, G. Wittum (eds.), *Multigrid Methods V.* Proceedings of the Fifth European Multigrid Conference held in Stuttgart, Germany, October 1-4, 1996. 1998. VIII, 334 pp. Softcover. ISBN 3-540-63133-X

Vol. 4   P. Deuflhard, J. Hermans, B. Leimkuhler, A. E. Mark, S. Reich, R. D. Skeel (eds.), *Computational Molecular Dynamics: Challenges, Methods, Ideas.* Proceedings of the 2nd International Symposium on Algorithms for Macromolecular Modelling, Berlin, May 21-24, 1997. 1998. XI, 489 pp. Softcover. ISBN 3-540-63242-5

Vol. 5   D. Kröner, M. Ohlberger, C. Rohde (eds.), *An Introduction to Recent Developments in Theory and Numerics for Conservation Laws.* Proceedings of the International School on Theory and Numerics for Conservation Laws, Freiburg / Littenweiler, October 20-24, 1997. 1998. VII, 285 pp. Softcover. ISBN 3-540-65081-4

Vol. 6   S. Turek, *Efficient Solvers for Incompressible Flow Problems.* An Algorithmic and Computational Approach. 1999. XVII, 352 pp, with CD-ROM. Hardcover. ISBN 3-540-65433-X

Vol. 7   R. von Schwerin, *Multi Body System SIMulation.* Numerical Methods, Algorithms, and Software. 1999. XX, 338 pp. Softcover. ISBN 3-540-65662-6

Vol. 8   H.-J. Bungartz, F. Durst, C. Zenger (eds.), *High Performance Scientific and Engineering Computing.* Proceedings of the International FORTWIHR Conference on HPSEC, Munich, March 16-18, 1998. 1999. X, 471 pp. Softcover. 3-540-65730-4

Vol. 9   T. J. Barth, H. Deconinck (eds.), *High-Order Methods for Computational Physics.* 1999. VII, 582 pp. Hardcover. 3-540-65893-9

Vol. 10   H. P. Langtangen, A. M. Bruaset, E. Quak (eds.), *Advances in Software Tools for Scientific Computing.* 2000. X, 357 pp. Softcover. 3-540-66557-9

Vol. 11   B. Cockburn, G. E. Karniadakis, C.-W. Shu (eds.), *Discontinuous Galerkin Methods.* Theory, Computation and Applications. 2000. XI, 470 pp. Hardcover. 3-540-66787-3

Vol. 12   U. van Rienen, *Numerical Methods in Computational Electrodynamics.* Linear Systems in Practical Applications. 2000. XIII, 375 pp. Softcover. 3-540-67629-5

Vol. 13   B. Engquist, L. Johnsson, M. Hammill, F. Short (eds.), *Simulation and Visualization on the Grid.* Parallelldatorcentrum Seventh Annual Conference, Stockholm, December 1999, Proceedings. 2000. XIII, 301 pp. Softcover. 3-540-67264-8

Vol. 14   E. Dick, K. Riemslagh, J. Vierendeels (eds.), *Multigrid Methods VI.* Proceedings of the Sixth European Multigrid Conference Held in Gent, Belgium, September 27-30, 1999. 2000. IX, 293 pp. Softcover. 3-540-67157-9

Vol. 15   A. Frommer, T. Lippert, B. Medeke, K. Schilling (eds.), *Numerical Challenges in Lattice Quantum Chromodynamics.* Joint Interdisciplinary Workshop of John von Neumann Institute for Computing, Jülich and Institute of Applied Computer Science, Wuppertal University, August 1999. 2000. VIII, 184 pp. Softcover. 3-540-67732-1

Vol. 16   J. Lang, *Adaptive Multilevel Solution of Nonlinear Parabolic PDE Systems.* Theory, Algorithm, and Applications. 2001. XII, 157 pp. Softcover. 3-540-67900-6

Vol. 17   B. I. Wohlmuth, *Discretization Methods and Iterative Solvers Based on Domain Decomposition.* 2001. X, 197 pp. Softcover. 3-540-41083-X

Vol. 18   U. van Rienen, M. Günther, D. Hecht (eds.), *Scientific Computing in Electrical Engineering.* Proceedings of the 3rd International Workshop, August 20-23, 2000, Warnemünde, Germany. 2001. XII, 428 pp. Softcover. 3-540-42173-4

Vol. 19   I. Babuška, P. G. Ciarlet, T. Miyoshi (eds.), *Mathematical Modeling and Numerical Simulation in Continuum Mechanics.* Proceedings of the International Symposium on Mathematical Modeling and Numerical Simulation in Continuum Mechanics, September 29 - October 3, 2000, Yamaguchi, Japan. 2002. VIII, 301 pp. Softcover. 3-540-42399-0

Vol. 20   T. J. Barth, T. Chan, R. Haimes (eds.), *Multiscale and Multiresolution Methods.* Theory and Applications. 2002. X, 389 pp. Softcover. 3-540-42420-2

Vol. 21   M. Breuer, F. Durst, C. Zenger (eds.), *High Performance Scientific and Engineering Computing.* Proceedings of the 3rd International FORTWIHR Conference on HPSEC, Erlangen, March 12-14, 2001. 2002. XIII, 408 pp. Softcover. 3-540-42946-8

Vol. 22   K. Urban, *Wavelets in Numerical Simulation.* Problem Adapted Construction and Applications. 2002. XV, 181 pp. Softcover. 3-540-43055-5

For further information on these books please have a look at our mathematics catalogue at the following URL: http://www.springer.de/math/index.html

Printing: Strauss GmbH, Mörlenbach
Binding: Schäffer, Grünstadt